MYWORKBOOK

CHRISTINE VERITY

PREALGEBRA

FIFTH EDITION

Margaret L. Lial

American River College

Diana L. Hestwood

Minneapolis Community and Technical College

Boston   Columbus   Indianapolis   New York   San Francisco   Upper Saddle River
Amsterdam   Cape Town   Dubai   London   Madrid   Milan   Munich   Paris   Montreal   Toronto
Delhi   Mexico City   São Paulo   Sydney   Hong Kong   Seoul   Singapore   Taipei   Tokyo

The author and publisher of this book have used their best efforts in preparing this book. These efforts include the development, research, and testing of the theories and programs to determine their effectiveness. The author and publisher make no warranty of any kind, expressed or implied, with regard to these programs or the documentation contained in this book. The author and publisher shall not be liable in any event for incidental or consequential damages in connection with, or arising out of, the furnishing, performance, or use of these programs.

Reproduced by Pearson from electronic files supplied by the author.

ISBN-13: 978-0-321-84522-1
ISBN-10: 0-321-84522-6

1 2 3 4 5 6 OPM 17 16 15 14 13

www.pearsonhighered.com

CONTENTS

Name: _____ Date: _____
Instructor: _____ Section: _____

Chapter 1 INTRODUCTION TO ALGEBRA: INTEGERS

1.1 Place Value

Learning Objectives
1   Identify whole numbers.
2   Identify the place value of a digit through hundred-trillions.
3   Write a whole number in words or digits.

Key Terms

Use the vocabulary terms listed below to complete each statement in exercises 1−3.

place value system   digits   whole numbers

1.   The ten _____ in our number system are 0, 1,2 3, 4, 5, 6, 7, 8, and 9.

2.   The _____ are 0, 1, 2, 3, 4, and so on.

3.   A _____ is a number system in which the location, or place, where a digit is written gives it a different value.

Guided Examples

Review this example for Objective 1:

1.   Identify the whole numbers in this list.

$$67, -3,\ 200,\ 1.9,\ \frac{8}{9},\ 0,\ 0.333,\ 8\frac{3}{4},\ 4$$

The whole numbers are 0, 4, 67, and 200.

Now Try:

1.   Identify the whole numbers in this list.

$$59, -6,\ 350,\ 9.7,\ \frac{9}{8},\ 0,\ 7\frac{1}{3},$$
$$0.999,\ 5$$

_____

Review this example for Objective 2:

2.   Identify the place of each 5 in the number. 7,946,145,057

7,946,145,057
           ↑ ↑ tens place
thousands place

Now Try:

2.   Identify the place of each 3 in the number.
      9,854,310,327

_____

Review these examples for Objective 3:	**Now Try:**

3.

   a. Write 7,096,381 in words.

   seven million, ninety-six thousand, three hundred eighty-one

   b. Write 41,677,000,500,000 in words.

   forty-one trillion, six hundred seventy-seven billion, five hundred thousand

4. Write each number using digits.

   a. Six hundred forty-two thousand, ten

   The answer is 642,010.

   b. Eighty-nine billion, twenty-five thousand, six hundred

   The answer is 89,000,025,600.

3.

   a. Write 9,075,862 in words.

   _____

   b. Write 54,800,543,700,100 in words.

   _____

4. Write each number using digits.

   a. Seven hundred fifty-three thousand, six

   _____

   b. Eleven billion, ten thousand, twenty-five

   _____

Objective 1   Identify whole numbers.

For extra help, see Example 1 on page 5 of your text and Section Lecture video for Section 1.1 and Exercise Solutions Clip 7.

Choose the whole numbers in each set of numbers.

1. $-\dfrac{3}{4}, 2, 1.398, -2\dfrac{1}{2}, -1.04, -2, \dfrac{1}{5}, -2.6428$

               1. _____

2. $-0.5, -1, 4, 3, -4.87, -22, \dfrac{4}{3}, 1\dfrac{1}{2}$

               2. _____

3. $2.718, -1, 365, 22.4, -6.02, 1, -4\dfrac{2}{5}$

               3. _____

Objective 2   Identify the place value of a digit through hundred-trillions.

For extra help, see Example 2 on page 5 of your text and Section Lecture video for Section 1.1 and Exercise Solutions Clip 9 and 15.

Give the place value of the digit 6 in each number.

4.   639,111,192

               4. _____

5. 94,164,372,757   5. _____

6. 6458   6. _____

Objective 3   Write a number in words or digits

For extra help, see Examples 3–4 on pages 5–6 of your text and Section Lecture video for Section 1.1 and Exercise Solutions Clip 21, 31, 35, and 39.

Write each number in words.

7. 59,504,806,873   7. _____

8. 9,671,000,637   8. _____

Write the number using digits.

9. Nine hundred eighty-seven million, three hundred   9. _____
thirty

Chapter 1 INTRODUCTION TO ALGEBRA: INTEGERS

1.2   Introduction to Integers

Learning Objectives
1   Write positive and negative numbers used in everyday situations.
2   Graph numbers on a number line.
3   Use the > and < symbols to compare integers.
4   Find the absolute value of integers.

Key Terms

Use the vocabulary terms listed below to complete each statement in exercises 1−3.

> number line   integers   absolute value

1.   The _____ of a number is its distance from 0 on the number line.

2.   A _____ is used to show how numbers relate to each other.

3.   The whole numbers together with their opposites and 0 are called _____.

Guided Examples

Review these examples for Objective 1:

1.   Write each negative number with a negative sign. Write each positive number in two ways.

   a. The Dead Sea is 1299 feet below sea level.

   The answer is –1299 ft.

   b. The mountain had elevation of 17,000 feet.

   The answer is +17,000 feet or 17,000 feet.

Now Try:

1.   Write each negative number with a negative sign. Write each positive number in two ways.
   a. A diver descends to a depth of 150 ft below sea level.

   _____

   b. The temperature rose to 68°F.

   _____

Name: Date:
Instructor: Section:

Review this example for Objective 2:

2. Graph each number on the number line.

$$-\frac{1}{2}, -3, -\frac{5}{2}, \frac{1}{4}, 1\frac{7}{8}, 3$$

Draw a dot at the correct location for each number.

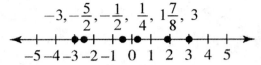

$$-3, -\frac{5}{2}, -\frac{1}{2}, \frac{1}{4}, 1\frac{7}{8}, 3$$

Now Try:

2. Graph each number on the number line.

$$-3, -5, -1\frac{1}{2}, \frac{2}{3}, 3.5, 4$$

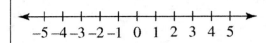

Review these examples for Objective 3:

3. Write < or > between each pair of integers to make a true statement.

   a. 0 _____ 7

   0 is to the left of 7 on the number line, so 0 is less than 7. Write 0 < 7.

   b. −2 _____ −5

   −2 is to the right of −5, so −2 is greater than −5. Write −2 > −5.

   c. −6 _____ 7

   −6 is to the left of 7, so −6 is less than 7. Write −6 < 7.

Now Try:

3. Write < or > between each pair of integers to make a true statement.

   a. 0 _____ −8

   _____

   b. −3 _____ −9

   _____

   c. 11 _____ −4

   _____

Review these examples for Objective 4:

4. Find each absolute value.

   a. $\left|3\right|$

   The distance from 0 to 3 on the number line is 3 spaces. So, $\left|3\right| = 3$.

   b. $\left|-3\right|$

   The distance from 0 to −3 on the number line is 3 spaces. So, $\left|-3\right| = 3$.

   c. $\left|0\right|$

   $\left|0\right| = 0$ because the distance from 0 to 0 on the number line is 0 spaces.

Now Try:

4. Find each absolute value.

   a. $\left|12\right|$

   _____

   b. $\left|-13\right|$

   _____

   c. $\left|0\right|$

   _____

Name: Date:
Instructor: Section:

Objective 1   Write positive and negative numbers used in everyday situations.

For extra help, see Example 1 on page 14 of your text and Section Lecture video for Section 1.2 and Exercise Solutions Clip 3 and 7.

Write a signed number for each of the following.

1.   Following a defeat, an army retreats 46 kilometers.   1. _____

2.   The company had a profit of $830.   2. _____

3.   A corporation has a shortfall of $1.2 million   3. _____

Objective 2   Graph numbers on a number line.

For extra help, see Example 2 on page 14 of your text and Section Lecture video for Section 1.2 and Exercise Solutions Clip 11.

Graph each set of numbers on a number line.

4. $-2, -1, 0, 1, 2, 5$

4.
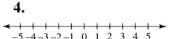

5. $-4.5, -1.5, -0.5, 0, 1.5, 2.5$

5.
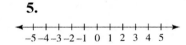

6. $4\frac{1}{2}, 1, -2\frac{3}{4}, -1\frac{1}{2}, -\frac{1}{4}$

6.

Objective 3   Use the < and > symbols to compare integers.

For extra help, see Example 3 on page 15 of your text and Section Lecture video for Section 1.2 and Exercise Solutions Clip 17, 23, and 31.

Write < or > in each blank to make a true statement.

7. -23 ____ -32 7. _____

8. -6 ____ 0 8. _____

9. -5 ____ -3 9. _____

7

Objective 4   Find the absolute value of integers.

For extra help, see Example 4 on page 16 of your text and Section Lecture video for Section 1.2 and Exercise Solutions Clip 33, 35, 41, 43, and 44.

Find each absolute value.

10. $|-95|$ 10. _____

11. $|21|$ 11. _____

12. $|-788|$ 12. _____

Chapter 1 INTRODUCTION TO ALGEBRA: INTEGERS

1.3   Adding Integers

Learning Objectives
1   Add integers.
2   Identify properties of addition.

Key Terms

Use the vocabulary terms listed below to complete each statement in exercises 1−5.

addends   **sum**   **addition property of 0**

commutative property of addition

associative property of addition

1.   By the_____, changing the order of the addends in an addition problem does not change the sum.

2.   In addition, the numbers being added are called the _____.

3.   By the_____, changing the grouping of the addends in an addition problem does not change the sum.

4.   The answer to an addition problem is called the _____.

5.   The _____ says that adding 0 to any number leaves the number unchanged.

Guided Examples

Review these examples for Objective 1:

1.   Use a number line to find $(-3)+(-2)$.

Start at 0. Move 3 places to the left. Then move 2 places to the left.

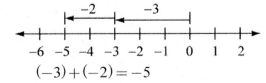

$(-3)+(-2)=-5$

Now Try:

1.   Use a number line to find $(-4)+(-1)$.

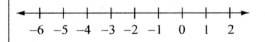

_____

2. Add.

a. $-9+(-4)$

Step 1 Add the absolute values.
$|-9|=9$ and $|-4|=4$
Add $9+4$ to get 13.
Step 2 Use the common sign as the sign of the sum. Both numbers are negative, so the sum is negative.
$-9+(-4)=-13$

b. $7+8$

Both numbers are positive, therefore the sum is positive.
$7+8=15$

3. Add.

a. $-7+4$

Step 1 $|-7|=7$ and $|4|=4$
Subtract $7-4$ to get 3.
Step 2 -7 has the greater absolute value and is negative, so the sum is also negative.
$-7+4=-3$

b. $-6+15$

Step 1 $|-6|=6$ and $|15|=15$
Subtract $15-6$ to get 9.
Step 2 15 has the greater absolute value and is positive, so the sum is also positive.
$-6+15=9$

4. A football team has to gain at least 10 yards during four plays in order to keep the ball. The football team lost 3 yards, gained 6 yards, lost 4 yards, and gained 9 yards. Did the team gain enough to keep the ball?

-3 yards $+6$ yards $+(-4$ yards$)+9$ yards

$\quad\quad$ 3 yards $\quad\quad +(-4$ yards$)+9$ yards

$\quad\quad\quad\quad -1$ yards $\quad\quad\quad +9$ yards

$\quad\quad\quad\quad\quad\quad$ 8 yards

The team gained 8 yards, which are not enough yards to keep the ball.

2. Add.

a. $(-25)+(-11)$

_____

b. $19+21$

_____

3. Add.

a. $-19+3$

_____

b. $-4+25$

_____

4. The temperature was $-20°$, then rose $32°$ and then dropped $15°$. What was the new temperature?

_____

Name: Date:

Instructor: Section:

Review these examples for Objective 2:

5. Rewrite each sum, using the commutative property of addition. Check that the sum is unchanged.

   a. $78 + 62$

   $78 + 62 = 62 + 78$

     $140 \quad = \quad 140$

   Both sums are 140, so the sum is unchanged.

   b. $-24 + (-53)$

   $-24 + (-53) = -53 + (-24)$

     $-77 \quad = \quad -77$

   Both sums are –77, so the sum is unchanged.

6. In each addition problem, pick out the two addends that would be easiest to add. Write parentheses around those addends. Then find the sum.

   a. $-10 + 10 + 7$

   Group $-10 + 10$ because the sum is 0.

     $(-10 + 10) + 7$

          $0 \quad + 7$

             7

   b. $-35 + 16 + 4$

   Group $16 + 4$ because the sum is 20, which is a multiple of 10.

     $-35 + (16 + 4)$

     $-35 + \quad 20$

       -15

Now Try:

5. Rewrite each sum, using the commutative property of addition. Check that the sum is unchanged.

   a. $254 + 63$

   b. $-44 + (-77)$

6. In each addition problem, pick out the two addends that would be easiest to add. Write parentheses around those addends. Then find the sum.

   a. $-20 + 19 + (-19)$

             _____

   b. $8 + 2 + (-17)$

             _____

Objective 1   Add integers.

For extra help, see Examples 1–4 on pages 21–23 of your text and Section Lecture video for Section 1.3 and Exercise Solutions Clip 3, 5, 7, 13, 21, and 23.

Add by using a number line.

   1. $-8 + 5$

          1. _____

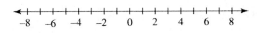

Add.

2. $7+(-22)$ 2.  _____

Write an addition problem for the situation and find the sum

3.   While playing a card game, Diane first gained 23   3. _____
         points, then lost 16 points, and finally gained 11
         points. What was her final score?

Objective 2   Identify properties of addition.

For extra help, see Examples 5–6 on pages 24–25 of your text and Section Lecture video
for Section 1.3 and Exercise Solutions Clip 71.

*Rewrite each sum using the commutative property of addition, and find the sum each
way.*

4. $-6+13$ 4. _____

*In each addition problem, write parentheses around the two addends that would be
easiest to add. Then find the sum.*

5. $8+\ ^-15+\ ^-5$ 5. _____

6. $^-11+11+2$ 6. _____

Chapter 1 INTRODUCTION TO ALGEBRA: INTEGERS

1.4   Subtracting Integers

Learning Objectives
1   Find the opposite of an integer.
2   Subtract integers.
3   Combine adding and subtracting of integers.

Key Terms

Use the vocabulary terms listed below to complete each statement in exercises 1–2.

opposite   **additive inverse**

1.   The _____of a number is the same distance from 0 on the
number line as the original number, but located on the other side of 0.

2.   The opposite of a number is its _____.

Guided Examples

Review these examples for Objective 1:

1.   Find the opposite (additive inverse) of each
number. Show that the sum of the number and its
opposite is 0.

   a. 4

   The opposite of 4 is –4 and $4+(-4)=0.$

   b. –15

   The opposite of –15 is 15 and $-15+15=0.$

   c. 0

   The opposite of 0 is 0 and $0+0=0.$

Now Try:

1.   Find the opposite (additive
inverse) of each number. Show
that the sum of the number and
its opposite is 0.

   a. 25

   _____

   b. –39

   _____

   c. 0

   _____

Review these examples for Objective 2:

2.   Make two pencil strokes to change each
subtraction problem into an addition problem.
Then find the sum.

   a. $6-9$

   Change 9 to –9. Change subtraction to addition.
   $6-9=6+(-9)=-3$

Now Try:

2.   Make two pencil strokes to
change each subtraction
problem into an addition
problem. Then find the sum.

   a. $13-15$

   _____

b. $-8-(-5)$

Change –5 to 5. Change subtraction to addition.
$-8-(-5)=-8+5=-3$

c. $4-(-9)$

Change –9 to 9. Change subtraction to addition.
$4-(-9)=4+9=13$

d. $-3-12$

Change 12 to –12. Change subtraction to addition.
$-3-12=-3+(-12)=-15$

b. $-10-(-9)$

_____

c. $12-(-11)$

_____

d. $-8-16$

_____

Review this example for Objective 3:

3. Simplify by completing all the calculations.
$-6-15-18+3$

Change all subtractions to adding the opposite. Change 15 to –15. Change 18 to –18. Then add from left to right.

$-6 - 15 - 18 + 3$

$-6+(-15)+(-18)+3$

$-21 + (-18)+3$

$-39 \qquad +3$

-36

Now Try:

3. Simplify by completing all the calculations.
$-9-14-21+6$

_____

Objective 1   Find the opposite of an integer.

For extra help, see Example 1 on page 30 of your text and Section Lecture video for Section 1.4 and Exercise Solutions Clip 1, 3, and 4.

Find the opposite (additive inverse) of each number. Show that the sum of the number and its opposite is 0.

1. -11

2. 4

3. -5

1. _____

   _____

2. _____

   _____

3. _____

   _____

Objective 2   Subtract integers.

For extra help, see Example 2 on page 31 of your text and Section Lecture video for Section 1.4 and Exercise Solutions Clip 9, 15, 17, 19, and 21.

Subtract by changing subtraction to addition.

4. $-2-(-11)$ 4. _____

5. $-13-1$ 5. _____

6. $12-(-5)$ 6. _____

Objective 3   Combine adding and subtracting of integers.

For extra help, see Example 3 on page 31 of your text and Section Lecture video for Section 1.4 and Exercise Solutions Clip 41.

Simplify.

7. $-2-16-(-18)$ 7. _____

8. $13-(-8)+(-7)-7$ 8. _____

9. $-14-(-3)-0+(-11)$ 9. _____

Chapter 1 INTRODUCTION TO ALGEBRA: INTEGERS

1.5   Problem Solving: Rounding and Estimating

Learning Objectives
1   Locate the place to which a number is to be rounded.
2   Round integers.
3   Use front end rounding to estimate answers in addition and subtraction.

Key Terms

Use the vocabulary terms listed below to complete each statement in exercises 1–3.

rounding   estimate   front end rounding

1. _____ is rounding to the highest possible place so that all the digits become zeros except the first one.

2. In order to find a number that is close to the original number, but easier to work with, use a process called _____.

3. _____ to find an answer close to the exact answer.

Guided Examples

Review these examples for Objective 1:

1. Locate and draw a line under the place to which each number is to be rounded. Then answer the question.

   a. Round –49 to the nearest ten. Is –49 closer to –40 or –50?

   –50 –49   –40

   –49 is closer to –50.

   b. Round $573 to the nearest hundred. Is $573 closer to $500 or $600?

   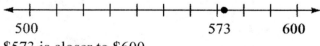

   500   573   600

   $573 is closer to $600.

   c. Round –63,155 to the nearest thousand. Is –63,155 closer to –63,000 or –64,000?

   –63,155 is closer to –63,000

Now Try:

1. Locate and draw a line under the place to which each number is to be rounded. Then answer the question.

   a. Round –32 to the nearest ten. Is –32 closer to –30 or –40?

   _____

   b. Round $642 to the nearest hundred. Is $642 closer to $600 or $700?

   _____

   c. Round –76,542 to the nearest thousand. Is –76,542 closer to –76,000 or –77,000?

   _____

Review this example for Objective 2:

2. Round 541 to the nearest hundred.

   Step 1 Locate the place to which the number is being rounded. Draw a line under that place.
   541
   hundred's place

   Step 2 Because the next digit to the right of the underlined place is 4, which is 4 or less, do not change the digit in the underlined place.
   541

   Step 3 Change all digits to the right of the underlined place to zeros.
   541 rounded to the nearest hundred is 500.

3. Round 67,899 to the nearest thousand.

   Step 1 Locate the place to which the number is being rounded. Draw a line under that place.
   67,899
   thousand's place

   Step 2 Because the next digit to the right of the underlined place is 8, which is 5 or more, add 1 to the underlined place.
   67,899
   change 7 to 8

   Step 3 Change all digits to the right of the underlined place to zeros.
   67,899 rounded to the nearest thousand is 68,000.

4.
   a. Round −5623 to the nearest ten.

   Step 1 −5623
   ten's place

   Step 2 The next digit to the right is 3, which is 4 or less.
   −5623
   Leave 2 as 2.

   Step 3 −5623 rounds to −5620.

   b. Round 37,968 to the nearest hundred.

   Step 1 37,968

Now Try:

2. Round 732 to the nearest hundred.

   _____

3. Round 94,567 to the nearest thousand.

   _____

4.
   a. Round −4738 to the nearest ten.

   _____

   b. Round 56,978 to the nearest hundred.

   _____

Step 2 The next digit to the right of 9 is 6, which is 5 or more.

    37,<u>9</u>68

        Change 9 to 10; write 0 and
        regroup 1 into thousands place.
        7 + regrouped 1 = 8

Step 3 37,968 rounds to 38,000.

5.

    a. Round –86,732 to the nearest ten-thousand.

Step 1 –<u>8</u>6,732

Step 2 The next digit to the right is 6, which is 5 or more.

    –<u>8</u>6,732

    Change 8 to 9.

Step 3 –<u>8</u>6,732 rounds to –90,000.

    b. Round 837,401,682 to the nearest million.

Step 1 93<u>7</u>,401,682

Step 2 The next digit is 4 or less.

    93<u>7</u>,401,682

    Leave 7 as 7.

Step 3 937,401,682 rounds to 937,000,000.

5.

    a. Round –75,493 to the nearest ten-thousand.

    _____

    b.

    _____

Review these examples for Objective 3:

6. Use front end rounding to round each number.

    a. –324

Round to the highest possible place, that is, the leftmost digit. In this case the leftmost digit, 3, is in the hundreds place, so round to the nearest hundred.

    For –<u>3</u>24, the next digit is 4 or less.
    –<u>3</u>24 rounds to –300.

    b. 95,602

The leftmost digit, 9, is in the ten-thousands place, so round to the nearest ten-thousand.

    For <u>9</u>5,602, the next digit is 5 or more.
    Change 9 to 10.
    Regroup 1 into millions place.
95,602 rounds to 1,000,000.

Now Try:

6. Use front end rounding to round each number.

    a. –849

    _____

    b. 974,312

    _____

7. Use front end rounding to estimate an answer. Then find the exact answer.
A checkbook had a balance of $912. A check was written for $394. What is the new balance?

Estimate: Use front end rounding to round $912 and $394.
    $912 rounds to $900 $394 rounds to $400
Use the rounded number and subtract to estimate the new checkbook balance.
    $900 – $400 = $500

Exact: Use the original numbers and subtract.
    $912 – $394 = $518

7. Use front end rounding to estimate an answer. Then find the exact answer.
A monthly gross pay is $3207, with deductions of $795. Find the net pay after deductions.

Estimate   _____

Exact   _____

Objective 1   Locate the place to which a number is to be rounded.

For extra help, see Example 1 on page 34 of your text and Section Lecture video for Section 1.5 and Exercise Solutions Clip 5 and 13.

Locate the place to which the number is rounded by writing the appropriate digit.

1.   ⁻135   Nearest ten   1. _____

2.   ⁻99,102   Nearest hundred   2. _____

3.   645,371   Nearest ten-thousand   3. _____

Objective 2   Round integers.

For extra help, see Examples 2–5 on pages 35–37 of your text and Section Lecture video for Section 1.5 and Exercise Solutions Clip 5 and 13.

Round each number to the indicated place.

4.   ⁻6694 to the nearest thousand   4. _____

5.   814,118 to the nearest ten thousand   5. _____

6.   ⁻4,697,134 to the nearest million   6. _____

Name: Date:
Instructor: Section:

Objective 3   Use front end rounding to estimate answers in addition and subtraction.

For extra help, see Examples 6–7 on pages 38–39 of your text and Section Lecture video for Section 1.5 and Exercise Solutions Clip 41 and 51.

Use front end rounding to round the number.

7.   The first printing run of a new book is 170,000
     copies.

7. _____

First use front end rounding to estimate each answer. Then find the exact answer.

8.   ⁻5172 − 3850

8.
Estimate_____

Exact _____

First use front end rounding to estimate the answer to the application problem. Then find the exact answer.

9.   A cannon is raised above the horizontal by an angle
     of 36°. The cannon is lowered 8°, then lowered
     another 17°, and finally raised 24°. What is the
     cannon's final elevation angle?

9.
Estimate_____

Exact _____

Chapter 1 INTRODUCTION TO ALGEBRA: INTEGERS

1.6   Multiplying Integers

Learning Objectives
1   Used a raised dot or parentheses to express multiplication.
2   Multiply integers.
3   Identify properties of multiplication.
4   Estimate answers to application problems involving multiplication.

Key Terms

Use the vocabulary terms listed below to complete each statement in exercises 1−7.

> **factors**   **product**   **multiplication property of 0**
>
> **multiplication property of 1**   **commutative property of multiplication**
>
> **associative property of multiplication**   **distributive property**

1. By the_____, changing the order of the factors in a multiplication problem does not change the product.

2. In multiplication, the numbers being multiplied are called the _____.

3. By the_____, changing the grouping of the factors in a multiplication problem does not change the product.

4. The _____ says that multiplying a number by 0 gives a product of 0.

5. The answer to a multiplication problem is called the _____.

6. The _____ states that $a(b + c) = ab + bc$.

7. The _____ says that multiplying a number by 1 leaves the number unchanged.

Guided Examples

Review these examples for Objective 1:

1. Rewrite each multiplication in three different ways, using a dot or parentheses. Also identify the factor and the product.

    a. 60×3

    Rewrite it as $60 \cdot 3$ or $60(3)$ or $(60)(3)$.
    The factors are 60 and 3. The product is 180.

Now Try:

1. Rewrite each multiplication in three different ways, using a dot or parentheses. Also identify the factor and the product.
    a. 80×6

    _____

b. 5×90

Rewrite it as $5 \cdot 90$ or $5(90)$ or $(5)(90)$.
The factors are 5 and 90. The product is 450.

b. 8×15

_____

Review these examples for Objective 2:

2. Multiply.

a. $-3 \cdot 9$

The factors have different signs, so the product is negative.
$$-3 \cdot 9 = -27$$

b. $-11(-7)$

The factors have the same sign, so the product is positive.
$$-11(-7) = 77$$

c. $8(-12)$

The factors have different signs, so the product is negative.
$$8(-12) = -96$$

3. Multiply.

a. $-7 \cdot (2 \cdot 4)$

Multiply inside the parentheses first.
$$-7 \cdot (2 \cdot 4)$$
$$-7 \cdot \ 8$$
$$-56$$
Note that the factors –7 and 8 have different signs, so the product is negative.

b. $-3 \cdot (-3) \cdot (-3)$

There is no work to do inside the parentheses, so multiply $-3 \cdot (-3)$ first. The factors have the same sign, so the product is positive.
$$-3 \cdot (-3) \cdot (-3)$$
$$9 \ \cdot (-3)$$
$$-27$$
Note that the factors 9 and –3 have different signs, so the product is negative.

Now Try:

2. Multiply.

a. $-6 \cdot 7$

_____

b. $-4(-15)$

_____

c. $9(-6)$

_____

3. Multiply.

a. $-5 \cdot (7 \cdot 2)$

_____

b. $-4 \cdot (-4) \cdot (-4)$

_____

Review these examples for Objective 3:

4. Multiply. Then name the property illustrated by each example.

   a. $(0)(-57)$

$(0)(-57) = 0$
This illustrates the multiplication property of 0.

   b. $763(1)$

$763(1) = 1$
This illustrates the multiplication property of 1.

5. Show that the product is unchanged and name the property that is illustrated in each case.

   a. $-5 \cdot (-8) = -8 \cdot (-5)$

$-5 \cdot (-8) = -8 \cdot (-5)$
$\quad 40 \quad = \quad 40$
This example illustrates the commutative property of multiplication.

   b. $6 \cdot (7 \cdot 3) = (6 \cdot 7) \cdot (3)$

$6 \cdot (7 \cdot 3) = (6 \cdot 7) \cdot (3)$
$\quad 6 \cdot 21 = 42 \cdot (3)$
$\quad 126 \quad = \quad 126$
This example illustrates the associative property of multiplication.

6. Rewrite each product using the distributive property. Show that the result is unchanged.

   a. $5(6+9)$

$5(6+9) = 5 \cdot 6 + 5 \cdot 9$
$\quad 5(15) \quad = 30 + 45$
$\quad 75 \quad = \quad 75$
Both results are 75.

   b. $-3(-8+2)$

$-3(-8+2) = (-3) \cdot (-8) + (-3) \cdot (2)$
$\quad -3(-6) \quad = \quad 24 \quad + \quad (-6)$
$\quad 18 \quad = \quad 18$
Both results are 18.

Now Try:

4. Multiply. Then name the property illustrated by each example.
   a. $671(0)$

   _____

   b. $1(-92)$

   _____

5. Show that the product is unchanged and name the property that is illustrated in each case.
   a. $-9 \cdot (-7) = -7 \cdot (-9)$

   _____

   b. $6 \cdot (7 \cdot 3) = (6 \cdot 7) \cdot (3)$

   _____

6. Rewrite each product using the distributive property. Show that the result is unchanged.
   a. $7(4+8)$

   _____

   b. $-9(-5+3)$

   _____

Review this example for Objective 4:

7. Use front end rounding to estimate an answer. Then find the exact answer.

A local cable company is losing about 1955 customers per month. How many customers are lost in a year?

Estimate: Use front end rounding: −1955 round to −2000, and 12 months rounds to 10 months.
$$(-2000) \cdot (10) = -20,000 \text{ customers}$$

Exact:
$$(-1955) \cdot (12) = -23,460 \text{ customers}$$

Now Try:

7. Use front end rounding to estimate an answer. Then find the exact answer.

Enrollment at a community college has decreased by 595 students for the last 3 semesters. What is the total decrease?

Estimate _____

Exact _____

Objective 1   Used a raised dot or parentheses to express multiplication.

For extra help, see Example 1 on page 44 of your text and Section Lecture video for Section 1.6 and Exercise Solutions Clip 5 and 29.

Rewrite each multiplication, first using a dot and then using parentheses.

1. $^-3 \times 6$

2. $12 \times {}^-3$

3. 4×5

1. _____

   _____

2. _____

   _____

3. _____

   _____

Objective 2   Multiply integers.

For extra help, see Examples 2–3 on page 46 of your text and Section Lecture video for Section 1.6 and Exercise Solutions Clip 11, 17, and 31.

Multiply.

4. $-7(-1)(-4)$

5. $-2 \cdot 31$

6. $-5 \cdot (-2) \cdot (6)$

4. _____

5. _____

6. _____

Objective 3   Identify properties of multiplication.

For extra help, see Examples 4–6 on pages 47–49 of your text and Section Lecture video for Section 1.6 and Exercise Solutions Clip 41, 49, 51, and 53.

Fill in the blank to make a true statement.

7. $0 = 18(\underline{\quad})$ 7. _____

Rewrite each multiplication, using the stated property. Show that the result is unchanged.

8.   Commutative property   8. _____
      $-22 \cdot 4$

9.   Associative property   9. _____
      $(3 \cdot (-2)) \cdot (-9)$

Objective 4   Estimate answers to application problems involving multiplication.

For extra help, see Example 7 on page 49 of your text and Section Lecture video for Section 1.6 and Exercise Solutions Clip 59.

First use front end rounding to estimate the answer to each application problem. Then find the exact answer.

10.   An airplane on an approach path to an airport is   **10.**
      losing altitude at the rate of 820 feet per minute.   **Estimate**_____
      How much altitude does the plane lose in 18
      minutes?   **Exact** _____

11.   A television is purchased for $375 down and twelve   **11.**
      monthly payments of $55 each. A set of speakers   **Estimate**_____
      costing a total of $280 is also part of the purchase.
      What is the total cost of the system?   **Exact** _____

12.   A tree grows from a seedling for 15 years at a rate of **12.**
      6 feet per year. At the end of that time, the tree is   **Estimate**_____
      felled with a cut 4 feet above the ground. Seven feet
      of the tip are trimmed off. How long is the resulting   **Exact** _____
      log?

Chapter 1 INTRODUCTION TO ALGEBRA: INTEGERS

1.7   Dividing Integers

Learning Objectives

1   Divide integers.
2   Identify properties of division.
3   Combine multiplying and dividing of integers.
4   Estimate answers to application problems involving division.
5   Interpret remainders in division application problems.

Key Terms

Use the vocabulary terms listed below to complete each statement in exercises 1–3.

      factors   **product**   **quotient**

1.   Numbers that are being multiplied are called _____.

2.   The answer to a division problem is called the _____.

3.   The answer to a multiplication problem is called the _____.

Guided Examples

Review these examples for Objective 1:

1.   Divide.

     a. $\dfrac{-30}{6}$

     The numbers have different signs, so the quotient is negative.

        $\dfrac{-30}{6} = -5$

     b. $\dfrac{-27}{-9}$

     The numbers have the same sign, so the quotient is positive.

        $\dfrac{-27}{-9} = 3$

     c. $70 \div (-5)$

     The numbers have different signs, so the quotient is negative.

        $70 \div (-5) = -14$

Now Try:

1.   Divide.

     a. $\dfrac{-35}{7}$

     _____

     b. $\dfrac{-40}{-10}$

     _____

     c. $24 \div (-8)$

     _____

Review these examples for Objective 2:

2. Divide. Then state the property illustrated by each example.

    a. $\dfrac{-672}{-672}$

    Any nonzero number divided by itself is 1.
    $$\dfrac{-672}{-672}=1$$

    b. $\dfrac{85}{1}$

    Any number divided by 1 is the number.
    $$\dfrac{85}{1}=85$$

    c. $\dfrac{0}{23}$

    Zero divided by any nonzero number is 0.
    $$\dfrac{0}{23}=0$$

    d. $\dfrac{67}{0}$

    Division by 0 is undefined.
    $\dfrac{67}{0}$ is undefined.

Now Try:

2. Divide. Then state the property illustrated by each example.

    a. $\dfrac{-703}{-703}$

    _____

    b. $\dfrac{93}{1}$

    _____

    c. $\dfrac{0}{39}$

    _____

    d. $\dfrac{51}{0}$

    _____

Review these examples for Objective 3:

3. Simplify.

    a. $8(-12)\div(-4\cdot 2)$

    $8(-12)\div(-4\cdot 2)$
    $8(-12)\div(-8)$
    $-96\div(-8)$
    12

    b. $-36\div(-3)(4)\div(-6)$

    $-36\div(-3)(4)\div(-6)$
    $12(4)\div(-6)$
    $48\div(-6)$
    -8

Now Try:

3. Simplify.

    a. $5(-16)\div(-2\cdot 4)$

    _____

    b. $-48\div(-4)(3)\div(-9)$

    _____

c. $-80 \div (-16) \div (-1)$	**c.** $-56 \div (-8) \div (-7)$
$-80 \div (-16) \div (-1)$	
$5 \div (-1)$	_____
-5	

Review this example for Objective 4:	**Now Try:**
4. Use front end rounding to estimate an answer. Then find the exact answer. Matthew lost 108 pounds last year. What was the average weight loss per month?	**4.** Use front end rounding to estimate an answer. Then find the exact answer. During the last year, John lost $3708 in the stock market. What was the average loss each month?
Estimate: Use front end rounding: -108 rounds to -100, and 12 months rounds to 10 months. $-100 \div 10 = -10$ pounds each month	
Exact: $-108 \div 12 = -9$ pounds each month	_____

Review these examples for Objective 5:	**Now Try:**
5. Divide; then interpret the remainder in each application.	**5.** Divide; then interpret the remainder in each application.
a. A reading club has $150 to buy children's books. If each book is $9, how many can be purchased? How many are left over?	**a.** John is serving 50 hot dogs at a party. If each package of rolls contains 8 rolls, how many packages should he buy?
We solve using division. $$\begin{array}{r} 16 \\ 9\overline{)150} \\ \underline{9} \\ 60 \\ \underline{54} \\ 6 \end{array}$$	_____
The club can buy 16 books. There will be $6 left over.	
b. The college marching band has 118 members and will spend the night in a hotel. Each room can accommodate 4 people. How many rooms are needed?	**b.** In the lobby of a skyscraper, 50 people are waiting for elevators. If the capacity of each elevator is 18 people, how many elevator trips are needed?
We use division to solve the problem. $$\begin{array}{r} 29 \\ 4\overline{)118} \\ \underline{8} \\ 38 \\ \underline{36} \\ 2 \end{array}$$	_____

If 29 rooms are rented, two members will have
to sleep on a bus. So, 30 rooms must be rented.
(One room will have only 2 band members.)

Objective 1   Divide integers.

For extra help, see Example 1 on page 54 of your text and Section Lecture video for
Section 1.7 and Exercise Solutions Clip 9, 15, 23, and 29.

Divide.

1. $\dfrac{28}{-4}$ 1. _____

2. $\dfrac{-36}{-12}$ 2. _____

3. $\dfrac{-100}{10}$ 3. _____

Objective 2   Identify properties of division.

For extra help, see Example 2 on page 55 of your text and Section Lecture video for
Section 1.7 and Exercise Solutions Clip 7, 17, and 27.

Complete the equation, then state the property represented by the equation.

4. $\dfrac{0}{14} =$ 4. _____

5. $\dfrac{-6}{-6} =$ 5. _____

6. $\dfrac{-48}{0} =$ 6. _____

Objective 3   Combine multiplying and dividing of integers.

For extra help, see Example 3 on page 56 of your text and Section Lecture video for
Section 1.7 and Exercise Solutions Clip 33 and 39.

Simplify.

7. $-120 \div (-6) \div 10$ 7. _____

8. $28 \cdot (2 \cdot (-1)) \div -8$ **8.** _____

9. $-84 \div (-3) \cdot 3 \div 4$ **9.** _____

Objective 4   Estimate answers to application problems involving division.

For extra help, see Example 4 on page 57 of your text and Section Lecture video for Section 1.7 and Exercise Solutions Clip 59.

Solve these application problems by using addition, subtraction, multiplication, or division. First estimate the answer using rounding; then find the exact answer.

10.   A honeybee flies from its hive a distance of 52 meters to one flowering bush, and from there it flies a distance of 17 meters from there to the next bush. What is the bee's total distance of travel so far?

10.
Estimate_____

Exact _____

11.   A car salesperson, bargaining with a customer over a car with a displayed price of $24,800, agrees to knock $2200 off that price. What is the new price at which the car is being offered?

11.
Estimate_____

Exact _____

12.   A sound wave has a frequency of 468 cycles per second. This frequency is multiplied by 5 to produce a new sound. What is the new frequency?

12.
Estimate_____

Exact _____

Name: Date:
Instructor: Section:

Objective 5   Interpreting Remainders in Division Applications.

For extra help, see Example 5 on pages 57–58 of your text and Section Lecture video for Section 1.7.

Divide, then interpret the remainder in each application.

13. A group of 96 construction workers is to be   13. _____
     transported on 7 trucks. The workers are to be
     divided as evenly as possible. How many workers
     will be transported on each truck?

14. A student plans to memorize 213 pages of notes in   14. _____
     seven days. How many pages per day should the
     student memorize?

15. A graduation ceremony will be held in an auditorium 15. _____
     with 1500 seats. If there are 468 students in the
     class, how many tickets will be allotted to each
     student?

Chapter 1 INTRODUCTION TO ALGEBRA: INTEGERS

1.8   Exponents and Order of Operations

Learning Objectives	
1	Use exponents to write repeated factors.
2	Simplify expressions containing exponents.
3	Use the order of operations.
4	Simplify expressions with fraction bars.

Key Terms

Use the vocabulary terms listed below to complete each statement in exercises 1−2.

      exponent   **order of operations**

1.   For problems or expressions with more than one operation, the
_____ tells what to do first, second, and so on, to obtain the correct answer.

2.   An _____ tells how many times a number is used as a factor
in repeated multiplication.

Guided Examples

Review these examples for Objective 1:

1. Given the factored form, give the exponential form, simplified form, and how it is read.

    a. $4 \cdot 4 \cdot 4$

    Exponential form: 4^3
    Simplified: 64
    Read as: 4 cubed, or 4 to the third power

    b. $(3)(3)$

    Exponential form: 3^2
    Simplified: 9
    Read as: 3 squared, or 3 to the second power

    c. 11

    Exponential form: 11^1
    Simplified: 11
    Read as: 11 to the first power

Now Try:

1. Given the factored form, give the exponential form, simplified form, and how it is read.

    a. $5 \cdot 5 \cdot 5 \cdot 5$

    _____

    b. $(6)(6)$

    _____

    c. 23

    _____

Name: Date:
Instructor: Section:

Review these examples for Objective 2:   **Now Try:**
2.   Simplify.   2.   Simplify.

   a. $(-4)^2$ **a.** $(-4)^2$

   $(-4)^2 = (-4)(-4) = 16$ _____

   b. $(-3)^3$ **b.** $(-10)^3$

   $(-3)^3 = (-3)(-3)(-3)$ _____
   $\qquad = 9(-3)$
   $\qquad = -27$

   c. $(-4)^4$ **c.** $(-3)^4$

   $(-4)^4 = (-4)(-4)(-4)(-4) = 256$ _____

   d. $5^3(-2)^2$ **d.** $4^2(-3)^3$

   $5^3(-2)^2 = \underbrace{(5)(5)(5)}\,\underbrace{(-2)(-2)}$ _____

   $\qquad = \;(125)\qquad(4)$
   $\qquad = 500$

Review these examples for Objective 3:   **Now Try:**
3.   Simplify.   3.   Simplify.

   a. $-9-(-7)+(-13)$ **a.** $-11-(-4)+(-19)$

   $-9-(-7)+(-13)$ _____
   $-9\;+7\;+(-13)$
   $\;-2\;+(-13)$
   $\quad -15$

   b. $-18 \div (-6)(5)$ **b.** $-27 \div (-9)(8)$

   $-18 \div (-6)(5)$ _____
   $\qquad (3)(5)$
   $\qquad 15$

4.   Simplify $8+5(24-6) \div 9$.   4.   Simplify $5+4(25-3) \div 11$.

   $8+5(24-6) \div 9$
   $8+\;5\;(18)\;\div 9$ _____
   $8\;+\;90\;\div 9$
   $\quad 8+10$
   $\qquad 18$

33

5. Simplify.

   a. $-9 \div (7-4) - 13$

   $-9 \div (7-4) - 13$

   $-9 \div (3) - 13$

   $-3 - 13$

   -16

   b. $5 + 3(9-14) \cdot (18 \div 6)$

   $5 + 3(9-14) \cdot (18 \div 6)$

   $5 + 3(-5) \cdot (18 \div 6)$

   $5 + 3(-5) \cdot (3)$

   $5 + (-15) \cdot (3)$

   $5 + (-45)$

   -40

6. Simplify.

   a. $6^2 - (-7)^2$

   $6^2 - (-7)^2$

   $6^2 - 49$

   $36 - 49$

   -13

   b. $(-3)^4 - (5-7)^2 (-9)$

   $(-3)^4 - (5-7)^2 (-9)$

   $(-3)^4 - (-2)^2 (-9)$

   $81 - 4(-9)$

   $81 - (-36)$

   $81 + 36$

   117

5. Simplify.

   a. $-10 \div (8-6) - 19$

   _____

   b. $6 + 4(11-15) \cdot (24 \div 8)$

   _____

6. Simplify.

   a. $8^2 - (-9)^2$

   _____

   b. $(-5)^3 - (6-9)^2 (-8)$

   _____

Review this example for Objective 4:

7.   Simplify $\dfrac{-9+6(7-10)}{5-6^2 \div 18}$.

    First do work in the numerator.

$$-9+6(7-10)$$

$$-9+\ 6(-3)$$

$$-9+(-18)$$

$$-27 \leftarrow \text{Numerator}$$

Now do the work in the denominator.

$$5-6^2 \div 18$$

$$5-36 \div 18$$

$$5-2$$

$$3 \leftarrow \text{Denominator}$$

The last step is the division.

$$\begin{array}{l}\text{Numerator} \rightarrow \\ \text{Denominator} \rightarrow\end{array} \dfrac{-27}{3} = -9$$

Now Try:

7.   Simplify $\dfrac{-15+7(9-12)}{8-10^2 \div 25}$.

_____

Objective 1   Use exponents to write repeated factors.

For extra help, see Example 1 on page 67 of your text and Section Lecture video for Section 1.8 and Exercise Solutions Clip 1, 5, and 7.

Rewrite each number in factored form as a number in exponential form; then state how the exponential form is read.

1. $12 \cdot 12 \cdot 12$

    1. _____

    _____

2. $6 \cdot 6 \cdot 6 \cdot 6 \cdot 6$

    2. _____

    _____

3. $7 \cdot 7 \cdot 7 \cdot 7 \cdot 7 \cdot 7 \cdot 7$

    3. _____

    _____

Objective 2   Simplify expressions containing exponents.

For extra help, see Example 2 on page 68 of your text and Section Lecture video for Section 1.8 and Exercise Solutions Clip 19, 25, and 27.

Simplify.

4. $(-1)^{91}$

    4. _____

5. $2^4 \cdot (-3)^2$ 5. _____

6. $(-2)^5 \cdot (-3)^2$ 6. _____

Objective 3   Use the order of operations.

For extra help, see Examples 3–6 on pages 68–71 of your text and Section Lecture video for Section 1.8 and Exercise Solutions Clip 45, 51, 55, and 67.

Simplify.

7. $3(-2+6)-(8-13)$ 7. _____

8. $3-(-6)\cdot(-1)^8$ 8. _____

9. $(-2)^3 \cdot (7-9)^2 \div 2$ 9. _____

Objective 4   Simplify expressions with fraction bars.

For extra help, see Example 7 on page 72 of your text and Section Lecture video for Section 1.8.

Simplify.

10. $\dfrac{-8+4^2-(-9)}{11-3-9}$ 10. _____

11. $\dfrac{-3\cdot 4^2 - 2(5+(-5))}{-3(8-13)\div -5}$ 11. _____

12. $\dfrac{4\cdot 2^2 - 8(7-2)}{7(6-8)\div 14}$ 12. _____

Chapter 2   UNDERSTANDING VARIABLES AND SOLVING EQUATIONS

2.1   Introduction to Variables

Learning Objectives

1	Identify variables, constants, and expressions.
2	Evaluate variable expressions for given replacement values.
3	Write properties of operations using variables.
4	Use exponents with variables.

Key Terms

Use the vocabulary terms listed below to complete each statement in exercises 1–5.

variable   constant   expression

evaluate the expression   coefficient

1.   An _____ tells the rule for doing something.

2.   A _____ is a letter that represents a number that varies or changes, depending on the situation.

3.   To _____, replace each variable with specific values and then follow the order of operations.

4.   A _____ is a number that is added or subtracted in an expression.

5.   The number part in a multiplication expression is the _____.

Guided Examples

Review this example for Objective 1:

1.   Write an expression for this rule. Identify the variable and the constant.
     Order the class limit minus 8 lunches because some students will brown bag.

     Let c represent the variable and 8 is the constant.
     $c - 8$

Now Try:

1.   Write an expression for this rule. Identify the variable and the constant.
     Order the class limit plus 5 extra lunches because some students are football players.

     _____

Review these examples for Objective 2:

2. Use this rule (from Example 1) for ordering lunches: Order the class limit minus 8. The expression is $c - 8$.

    a. Evaluate the expression when the class limit is 39.

    Replace c with 39 and follow the rule.

$$c - 8$$
$$39 - 8$$
$$31 \quad \text{Order 31 lunches.}$$

    b. Evaluate the expression when the class limit is 22.

    Replace c with 22 and follow the rule.

$$c - 8$$
$$22 - 8$$
$$14 \quad \text{Order 14 lunches.}$$

3. The expression (rule) for finding the perimeter of a square shape is $4s$. Evaluate the expression when the length of a side of a square garden is 20 feet.

    Replace s with 20 feet.

    The total distance around the square garden is 80 feet.

4. Evaluate the expression $100 + \dfrac{a}{2}$ when the age of the person is 36.

    Replace a with 36, the age of the person. Then follow the rule using the order of operations.

$$100 + \frac{a}{2}$$
$$100 + \frac{36}{2}$$
$$100 + 18$$
$$118$$

The approximate systolic blood pressure is 118.

Now Try:

2. Use this rule (from Now Try 1) for ordering lunches: Order the class limit plus 8. The expression is $c + 5$.

    a. Evaluate the expression when the class limit is 24.

    _____

    b. Evaluate the expression when the class limit is 45.

    _____

3. The expression (rule) for finding the perimeter of a square shape is $4s$. Evaluate the expression when the length of a side of a square garden is 8 yards.

    _____

4. Evaluate the expression $100 + \dfrac{a}{2}$ when the age of the person is 56.

    _____

5.

a. Find your average score if you bowl three games and your total score for all three games is 384.

Use the expression (rule) for finding your average score. Replace t with your total score of 384, and replace g with 3, the number of games.

$$\frac{t}{g} \begin{array}{c}\to 384 \\ \to \ \ 3\end{array}$$

Divide 384 by 3 to get 128. Your average score is 128.

b. Complete these tables to show how to evaluate each expression.

Value of x	Value of y	Expression (Rule) $x+y$
3	6	$3+6$ is 9
-7	4	__ $+$ __ is __
0	17	__ $+$ __ is __

Value of x	Value of y	Expression (Rule) xy
3	4	$3\cdot4$ is 12
-8	3	__ $\cdot$ __ is __
0	19	__ $\cdot$ __ is __

The expression (rule) is to *add* the two variables. So the completed table is:

Value of x	Value of y	Expression (Rule) $x+y$
3	6	$3+6$ is 9
-7	4	$-7+4$ is -3
0	17	$0+17$ is 17

The expression (rule) is to *multiply* the two variables. So the completed table is:

Value of x	Value of y	Expression (Rule) xy
3	4	$3\cdot4$ is 12
-8	3	$-8\cdot3$ is -24
0	19	$0\cdot19$ is 0

5.

a. Find your average score if you bowl four games and your total score for all four games is 384.

_____

b. Complete these tables to show how to evaluate each expression.

Value of x	Value of y	Expression (Rule) $x+y$
10	21	__ $+$ __ is __
-12	9	__ $+$ __ is __
93	0	__ $+$ __ is __

Value of x	Value of y	Expression (Rule) xy
7	6	__ $\cdot$ __ is __
-9	5	__ $\cdot$ __ is __
31	0	__ $\cdot$ __ is __

Review this example for Objective 3:	**Now Try:**

Review this example for Objective 3:

6. Use the variable b to state the property: When any number is added to 0, the sum is the number.

Use the letter b to represent any number.
$$b + 0 = b$$

Now Try:

6. Use the variable b to state the property: When any non-zero number is divided into 0, the quotient is zero.

_____

Review these examples for Objective 4:

7. Rewrite each expression without exponents.

a. x^6

x^6 can be written as $x \cdot x \cdot x \cdot x \cdot x \cdot x$.

b. $13a^2b$

$13a^2b$ can be written as $13 \cdot a \cdot a \cdot b$.

c. $-10p^3q^4$

$-10p^3q^4$ can be written as $-10 \cdot p \cdot p \cdot p \cdot q \cdot q \cdot q \cdot q$.

8. Evaluate each expression.

a. x^3 when x is -5

x^3 means $x \cdot x \cdot x$ Replace each x with -5.
$$(-5) \cdot (-5) \cdot (-5)$$
$$25 \quad \cdot (-5)$$
$$-75$$
So x^3 becomes $(-5)^3$, which is $(-5) \cdot (-5) \cdot (-5)$, or -125.

b. x^2y^3 when x is -6 and y is -3

Replace x with -6 and replace y with -3.
x^2y^3 means $x \cdot x \cdot y \cdot y \cdot y$
$$(-6) \cdot (-6) \cdot (-3) \cdot (-3) \cdot (-3)$$
$$36 \quad \cdot (-3) \cdot (-3) \cdot (-3)$$
$$-108 \quad \cdot (-3) \cdot (-3)$$
$$324 \quad \cdot (-3)$$
$$-972$$
So x^2y^3 becomes $(-6)^2(-3)^3$, which is $(-6) \cdot (-6) \cdot (-3) \cdot (-3) \cdot (-3)$, or -972.

Now Try:

7. Rewrite each expression without exponents.

a. z^4

_____

b. $14rs^2$

_____

c. $-13m^3n^5$

_____

8. Evaluate each expression.

a. z^2 when z is -9

_____

b. x^4y^2 when x is -2 and y is -3

_____

c. $-7a^2b$ when a is 4 and b is 6

Replace a with 4 and replace b with 6.

$-7a^2b$ means $-7 \cdot a \cdot a \cdot b$

$$-7 \cdot 4 \cdot 4 \cdot 6$$
$$-28 \cdot 4 \cdot 6$$
$$-112 \cdot 6$$
$$-672$$

So $-7a^2b$ becomes $-7(4)^2(6)$, which is $-7 \cdot 4 \cdot 4 \cdot 6$, or -672.

c. $-6c^3d$ when c is 2 and d is 5

_____

Objective 1   **Identify variables, constants, and expressions.**

For extra help, see Example 1 on page 95 of your text and Section Lecture video for Section 2.1.

Identify the parts of each expression. Choose from **variable**, **constant**, *and* **coefficient**.

1. $-7 + h$

             1. _____

             _____

2. $-2w$

             2. _____

             _____

3. $9k + 1$

             3. _____

             _____

Objective 2   **Evaluate variable expressions for given replacement values.**

For extra help, see Examples 2–5 on pages 95–98 of your text and Section Lecture video for Section 2.1 and Exercise Solutions Clip 17 and 23.

Evaluate each expression.

4.   The expression (rule) for the weight of an object (in Newtons) is $10m$, where m is the object's mass in kilograms. Evaluate the expression when
   (a) the object has a mass of 14 kilograms.
   (b) the object has a mass of 92 kilograms.

     4. **a.**_____

         b._____

5. The expression (rule) for the total number of dollars
    spent on a car, when there is a $2300 down payment
    and monthly payments of $215, is $2300 + 215t$,
    where t is the number of monthly payments.
    Evaluate the expression when
    (a) there are 36 monthly payments.
    (b) there are 48 monthly payments.

5. **a.**_____

  b._____

6. The expression (rule) for the amount of insurance
    paid out on a policy with a $2500 deductible is
    $\frac{8c}{10} - \$2500$, where c is total medical costs. Evaluate
    the expression when
    (a) total medical costs are $10,625.
    (b) total medical costs are $12,480.

6. **a.**_____

  b._____

Objective 3   Write properties of operations using variables.

For extra help, see Example 6 on page 99 of your text and Section Lecture video for
Section 2.1.

State the property represented by each equation.

7. $h(a + b) = h(a) + h(b)$

7. _____

8. $(1)(k) = k$

8. _____

9. $x + 0 = x$

9. _____

Objective 4   Use exponents with variables.

For extra help, see Examples 7–8 on page 100 of your text and Section Lecture video for Section 2.1 and Exercise Solutions Clip 35, 41, and 49.

Rewrite each expression without exponents.

10. $-7g^3h$

10. _____

11. $u^3v^2w^2$

11. _____

Evaluate the expression.

12. $-3k^2g^3$ when k is -2 and g is 5

12. _____

Chapter 2 UNDERSTANDING VARIABLES AND SOLVING EQUATIONS

2.2   Simplifying Expressions

Learning Objectives
1   Combine like terms using the distributive property.
2   Simplify expressions.
3   Use the distributive property to multiply.

Key Terms

Use the vocabulary terms listed below to complete each statement in exercises 1–5.

simplify an expression   **term**   **constant term**

variable term   **like terms**

1. _____are terms with exactly the same variable parts.

2. Each addend in an expression is called a _____.

3. To _____, combine all the like terms.

4. A _____ has a coefficient multiplied by a variable part.

5. A _____ is a type of term that is just a number.

Guided Examples

Review these examples for Objective 1:
1. List the like terms in each expression. Then identify the coefficients of the like terms.

   a. $-6w+(-6w^2)+4wz+w+(-6)$

   The like terms are $-6w$ and w.
   The coefficient of $-6w$ is -6, and the coefficient of w is understood to be 1.

   b. $9ab^2+9a^2b+(-4a^2b)+3+(-9ab)$

   The like terms are $9a^2b$ and $-4a^2b$.
   The coefficient of $9a^2b$ is 9, and the coefficient of $-4a^2b$ is -4.

Now Try:
1. List the like terms in each expression. Then identify the coefficients of the like terms.

   a. $-8a+(-8a^2)+9ab$
   $+a+(-11)$

   _____

   _____

   b. $21rs^2+21r^2s+(-5r^2s)$
   $+11+(-33rs)$

   _____

   _____

c. $12xy + 19 + (-18x) + 15y + (-7)$

The like terms are 19 and –7.
The like terms are constants (there are no variable parts).

c. $13pq + 26 + (-10p)$
$+15q + (-2)$

_____

_____

2. Combine like terms.

a. $7x + x + 9x$

$7x + x + 9x$
$7x + 1x + 9x$
$(7 + 1 + 9)x$
$17x$

2. Combine like terms.

a. $12x + 7x + x$

_____

b. $-5y^2 - 9y^2$

$-5y^2 - 9y^2$
$-5y^2 + (-9y^2)$
$[-5 + (-9)]y^2$
$-14y^2$

b. $-13z^2 - 26z^2$

_____

Review these examples for Objective 2:

3. Simplify each expression by combining like terms.

a. $9ab + 7b + 8ab$

The like terms are $9ab$ and $8ab$. Use the commutative property first, then combine like terms.

$9ab + 7b + 8ab$
$9ab + 8ab + 7b$
$(9 + 8)ab + 7b$
$17ab \quad + 7b$

The simplified expression is $17ab + 7b$.

Now Try:

3. Simplify each expression by combining like terms.

a. $5pq + q + 6pq$

_____

b. $4c - 9 - c + 11$

Write 1 as the coefficient of c. Change subtractions to adding the opposite, and then combine like terms.

$4c \ - \ 9 \ - \ 1c \ +11$

$4c + (-9) + (-1c) + 11$

$4c + (-1c) + (-9) + 11$

$[4 + (-1)]c + (-9) + 11$

$\quad 3c \ \ + \ \ 2$

The simplified expression is $3c + 2$.

b. $8c - 13 - c + 4$

_____

4. Simplify.

a. $6(11x)$

Use the associative property.

$\quad 6(11x)$ can be written as $(6 \cdot 11) \cdot x$

$\qquad\qquad\qquad 66 \cdot x$

$\qquad\qquad\qquad 66x$

So, $6(11x)$ simplifies to $66x$.

b. $-7(5b)$

Use the associative property.

$\quad -7(5b)$ can be written as $(-7 \cdot 5)b$

$\qquad\qquad\qquad -35b$

So, $-7(5b)$ simplifies to $-35b$.

c. $-9(-3a^2)$

Use the associative property.

$\quad -9(-3a^2)$ can be written as $[-9 \cdot (-3)]a^2$

$\qquad\qquad\qquad 27a^2$

So, $-9(-3a^2)$ simplifies to $27a^2$.

4. Simplify.

a. $9(12b)$

_____

b. $-4(9x)$

_____

c. $-8(-5y^2)$

_____

Review these examples for Objective 3:

5. Simplify.

a. $7(4x + 3)$

$7(4x + 3)$ can be written as $7 \cdot 4x + 7 \cdot 3$

$\qquad\qquad\qquad\qquad 7 \cdot 4 \cdot x + 21$

$\qquad\qquad\qquad\qquad 28 \cdot x + 21$

$\qquad\qquad\qquad\qquad 28x + 21$

So, $7(4x + 3)$ simplifies as $28x + 21$.

Now Try:

5. Simplify.

a. $6(5x + 2)$

_____

b. $-5(8b+4)$

$-5(8b+4)$ can be written as $-5 \cdot 8b + (-5) \cdot 4$

$$-5 \cdot 8 \cdot b + (-20)$$
$$-40 \cdot b + (-20)$$
$$-40b + (-20)$$

So, $-5(8b+4)$ simplifies as $-40b - 20$.

c. $8(d-9)$

$8(d-9)$ can be written as $8[d + (-9)]$

$$8 \cdot d + 8(-9)$$
$$8d + (-72)$$
$$8d - 72$$

So, $8(d-9)$ simplifies as $8d - 72$.

6. Simplify: $9 + 4(x-5)$

Use the distributive property first.
$$9 + 4(x-5)$$
$$9 + 4 \cdot x - 4 \cdot 5$$
$$9 + 4x - 20$$
$$4x + 9 - 20$$
$$4x \quad -11$$

The simplified expression is $4x - 11$.

b. $-9(6x+4)$

_____

c. $28(x-4)$

_____

6. Simplify: $12 + 6(x-3)$

_____

Objective 1   Combine like terms using the distributive property.

For extra help, see Examples 1–2 on pages 108–109 of your text and Section Lecture video for Section 2.2 and Exercise Solutions Clip 19.

Identify the like terms in each expression. Then identify the coefficients of the like terms.

1. $x^2 + 2x + (-5x^3) + (-3x) + 8$

1. _____

_____

Simplify each expression.

2. $6d + d$

2. _____

3. $c^3 z^4 - 11c^3 z^4 - 9c^3 z^4$

3. _____

Objective 2   Simplify expressions.

For extra help, see Examples 3–4 on pages 110–111 of your text and Section Lecture video for Section 2.2 and Exercise Solutions Clip 31 and 47.

Simplify each expression by combining like terms. Write each answer with the variables in alphabetical order and any constant term last.

4. $5b^4m + 10 - 2b^4m - 1$

4. _____

5. $-8s + 6st - 9t - 11st + 2s + 4t + 9 - 11t - 1$

5. _____

Simplify by using the associative property of multiplication.

6. $-14(-a^3c^2t)$

6. _____

Objective 3   Use the distributive property to multiply.

For extra help, see Examples 5–6 on pages 112–113 of your text and Section Lecture video for Section 2.2 and Exercise Solutions Clip 53, 55, 57, and 67.

Use the distributive property to simplify each expression.

7. $-5(t + 4)$

7. _____

8. $-8(3s - 4)$

8. _____

Simplify the expression.

9. $-q + 7(3q - 2) + 11$

9. _____

Chapter 2 UNDERSTANDING VARIABLES AND SOLVING EQUATIONS

2.3   Solving Equations Using Addition

Learning Objectives
1   Determine whether a given number is a solution of an equation.
2   Solve equations using the addition property of equality.
3   Simplify equations before using the addition property of equality.

Key Terms

Use the vocabulary terms listed below to complete each statement in exercises 1−5.

equation   **solve an equation**   **solution**

addition property of equality   **check the solution**

1.   A_____ of an equation is a number that makes the statement true when it is substituted for the variable.

2.   The _____ states that adding the same quantity to both sides of an equation keeps the equation balanced.

3.   An _____ is a mathematical statement that contains an equals sign.

4.   To _____, find a number that can replace the variable and make the equation balance.

5.   To _____, go back to the original equation and replace the variable with the solution.

Guided Examples

Review this example for Objective 1:

1.   Which of these numbers, 65, 45, or 85, is the solution of the equation $c - 15 = 70$?

Replace c with each of the numbers. The one that makes the equation balance is the solution.

$65 - 15 \neq 70$ Does not balance: $65 - 15 = 50$.

$45 - 15 \neq 70$ Does not balance: $45 - 15 = 30$.

$85 - 15 = 70$ Balances: $85 - 15 = 70$.

The solution is 85 because when c is 85, the equation balances.

Now Try:

1.   Which of these numbers, 25, 29, or 21, is the solution of the equation $25 = c - 4$?

_____

Review these examples for Objective 2:

2. Solve each equation and check the solution.

    a. $c + 7 = 20$

    Add -7 to each side.
$$c + 7 = 20$$
$$\underline{-7 \quad -7}$$
$$c + 0 = 13$$
$$c = 13$$

    Because c balances with 13, the solution is 13.

    Check the solution by replacing c with 13 in the original equation.
$$c + 7 = 20$$
$$13 + 7 = 20$$
$$20 = 20$$

    The equation balances. Therefore, 13 is the correct solution.

    b. $-7 = x - 4$

    Change subtraction to adding the opposite. Then Add the opposite of -4, which is 4.
$$-7 = x - 4$$
$$-7 = x + (-4)$$
$$\underline{ 4 \qquad\quad 4}$$
$$-3 = x + 0$$
$$-3 = x$$

    Because x balances with -3, the solution is -3.

    Check the solution by replacing c with -3 in the original equation.
$$-7 = x - 4$$
$$-7 = -3 - 4$$
$$-7 = -3 + (-4)$$
$$-7 = -7$$

    When x is replaced with -3, the equation balances, so -3 is the correct solution.

Now Try:

2. Solve each equation and check the solution.

    a. $c + 13 = 29$

    _____

    b. $-10 = x - 8$

    _____

Name:

Date:

Instructor:

Section:

Review these examples for Objective 3:

3. Solve each equation and check the solution.

   a. $y + 9 = 5 - 8$

   Simplify the right side first. Then get y by itself on the left side by adding the opposite of 9, which is -9.

   $$y + 9 = 5 - 8$$
   $$y + 9 = 5 + (-8)$$
   $$y + 9 = -3$$
   $$\underline{-9 \quad -9}$$
   $$y = -12$$

   The solution is -12. Now check the solution.

   Check   Go back to the original equation and replace y with -12.

   $$y + 9 = 5 - 8$$
   $$-12 + 9 = 5 - 8$$
   $$-3 = 5 + (-8)$$
   $$-3 = -3$$

   So when y is replaced with -12, the equation balances, so -12 is the correct solution.

   b. $-5 + 5 = -7b - 10 + 8b$

   Combine like terms on each side first.

   $$-5 + 5 = -7b - 10 + 8b$$
   $$0 = -7b + (-10) + 8b$$
   $$0 = -7b + 8b + (-10)$$
   $$0 = 1b + (-10)$$
   $$\underline{10 10}$$
   $$10 = b$$

   The solution is 10.

   Check Go back to the original equation and replace b with 10.

   $$-5 + 5 = -7b - 10 + 8b$$
   $$-5 + 5 = -7 \cdot 10 + (-10) + 8 \cdot 10$$
   $$0 = -70 + (-10) + 80$$
   $$0 = -80 + 80$$
   $$0 = 0$$

   When b is replaced with 10, the equation balances, so 10 is the correct solution.

Now Try:

3. Solve each equation and check the solution.

   a. $x + 10 = 14 - 15$

   _____

   b. $-9 + 9 = -15b - 6 + 16b$

   _____

Objective 1   Determine whether a given number is a solution of an equation.

For extra help, see Example 1 on page 123 of your text and Section Lecture video for Section 2.3 and Exercise Solutions Clip 3.

In each list of numbers, find the one that is a solution of the given equation.

1. $r - 12 = 0$ 1. _____

        $-8, -2, \ 12$

2. $p + 7 = 10$ 2. _____

        $-7, 3, 7$

3. $6 - 2y = -8$ 3. _____

        $-3, 5, 7$

Objective 2   Solve equations using the addition property of equality.

For extra help, see Example 2 on pages 125–126 of your text and Section Lecture video for Section 2.3 and Exercise Solutions Clip 11.

Solve each equation and check the solution.

4. $v - 72 = 32$ 4. _____

5. $-30 = 36 + n$ 5. _____

6. $65 = u + 74$ 6. _____

Name: Date:

Instructor: Section:

Objective 3   Simplify equations before using the addition property of equality.

For extra help, see Example 3 on page 126–127 of your text and Section Lecture video for Section 2.3 and Exercise Solutions Clip 39 and 43.

Simplify each side of the equation when possible. Then solve the equation. Check each solution.

7. $11a - 10a = -5 + 8$ 7. _____

8. $40 - 2u + 3u = 19$ 8. _____

9. $66n + 53 - 65n = 9$ 9. _____

Chapter 2 UNDERSTANDING VARIABLES AND SOLVING EQUATIONS

2.4   Solving Equations Using Division

Learning Objectives
1   Solve equations using the division property of equality.
2   Simplify equations before using the division property of equality.
3   Solve equations such as $-x = 5$.

Key Terms

Use the vocabulary terms listed below to complete each statement in exercises 1–2.

division property of equality

addition property of equality

1. The _____ states that dividing both sides of an equation by the same nonzero number will keep it balanced.

2. When the same quantity is added to both sides of an equation, the _____ is being applied.

Guided Examples

Review these examples for Objective 1:
1. Solve each equation and check each solution.

   a. $5s = 35$

   On the left side of the equation, the variable is multiplied by 5. To undo the multiplication, divide by 5.
   $$5s = 35$$
   $$\frac{5s}{5} = \frac{35}{5}$$
   $$s = 7$$
   So, 7 is the solution. We check the solution by replacing s with 7 in the original equation.

   Check
   $$5s = 35$$
   $$5 \cdot 7 = 35$$
   $$35 = 35$$
   When s is replaced with 7, the equation balances, so 7 is the correct solution.

Now Try:
1. Solve each equation and check each solution.

   a. $9s = 54$

   _____

b. $56 = -7w$

On the right side of the equation, the variable is multiplied by –7. To undo the multiplication, divide by –7.

$$56 = -7w$$

$$\frac{56}{-7} = \frac{-7w}{-7}$$

$$-8 = w$$

The solution is –8.

Check
We check the solution in the original equation.

$$56 = -7w$$

$$56 = -7 \cdot (-8)$$

$$56 = 56$$

When w is replaced with –8, the equation balances, so –8 is the correct solution.

b. $64 = -4w$

_____

Review these examples for Objective 2:

2. Solve each equation and check each solution.

a. $6y - 10y = -20$

First, simplify the left side by combining like terms.

$$6y - 10y = -20$$

$$6y + (-10y) = -20$$

$$\frac{-4y}{-4} = \frac{-20}{-4}$$

$$y = 5$$

The solution is 5.

Check
Go back to the original equation and replace each y with 5.

$$6y - 10y = -20$$

$$6 \cdot 5 - 10 \cdot 5 = -20$$

$$30 - 50 = -20$$

$$30 + (-50) = -20$$

$$-20 = -20$$

When y is replaced with 5, the equation balances, so 5 is the correct solution.

Now Try:

2. Solve each equation and check each solution.

a. $5y - 14y = -36$

_____

b. $13 - 21 + 8 = h + 6h$

First, combine like terms.

$$13 - 21 + 8 = h + 6h$$

$$13 + (-21) + 8 = 1h + 6h$$

$$-8 \quad + 8 = 7h$$

$$\frac{0}{7} = \frac{7h}{7}$$

$$0 = h$$

The solution is 0.

Check Go back to the original equation and replace each h with 0.

$$13 - 21 + 8 = h + 6h$$

$$13 + (-21) + 8 = 0 + 6 \cdot 0$$

$$-8 \quad + 8 = 0 + 0$$

$$0 = 0$$

When h is replaced with 0, the equation balances, so 0 is the correct solution.

b. $24 - 43 + 19 = h - 9h$

_____

Review this example for Objective 3:

3. Solve $-x = 9$ and check the solution.

Start by writing in the coefficient of x.

$$-x = 9$$

$$-1x = 9$$

$$\frac{-1x}{-1} = \frac{9}{-1}$$

$$x = -9$$

The solution is –9.

Check Go back to the original equation and replace x with 0.

$$-x = 9$$

$$-(-9) = 9$$

$$9 = 9$$

When x is replaced with –9, the equation balances, so –9 is the correct solution.

Now Try:

3. Solve $-y = 13$ and check the solution.

_____

Objective 1   **Solve equations using the division property of equality.**

For extra help, see Example 1 on page 135 of your text and Section Lecture video for Section 2.4 and Exercise Solutions Clip 9.

Solve each equation and check.

   1. $-3m = 21$ **1.** _____

   2. $-56 = -7b$ **2.** _____

   3. $15h = 420$ **3.** _____

Objective 2   **Simplify equations before using the division property of equality.**

For extra help, see Example 2 on page 136 of your text and Section Lecture video for Section 2.4 and Exercise Solutions Clip 23 and 27.

Solve each equation and check.

   4. $53 - 8 = -9d$ **4.** _____

   5. $35m - 33m = 100 - 34$ **5.** _____

6. $140 - 116 = 20x - 12x$ **6.** _____

Objective 3   **Solve equations such as** $-x = 5$**.**

For extra help, see Example 3 on page 137 of your text and Section Lecture video for Section 2.4 and Exercise Solutions Clip 39.

Solve each equation and check.

7. $-b = 19$ **7.** _____

8. $-17 = -w$ **8.** _____

9. $4 = -v$ **9.** _____

Chapter 2 UNDERSTANDING VARIABLES AND SOLVING EQUATIONS

2.5   Solving Equations with Several Steps

Learning Objectives
1   Solve equations using the addition and division properties of equality.
2   Solve equations using the distributive, addition, and division properties.

Key Terms

Use the vocabulary terms listed below to complete each statement in exercises 1–3.

distributive property

division property of equality

addition property of equality

1.   The _____ states that if $a = b$, then $a + c = b + c$.

2.   The _____ states that $a(b + c) = ab + bc$.

3.   The _____ states that if $a = b$, then $\dfrac{a}{c} = \dfrac{b}{c}$ for $c \neq 0$.

Guided Examples

Review these examples for Objective 1:

1.   Solve this equation and check the solution: $7m + 2 = 37$.

Step 1 Adding –2 to the left side will leave $7m$ by itself. To keep the balance, add –2 to the right side also.

$$7m + 2 = 37$$
$$\underline{-2 \quad -2}$$
$$7m + 0 = 35$$
$$7m = 35$$

Step 2 Divide both sides by the coefficient of the variable term, 7.

$$\frac{7m}{7} = \frac{35}{7}$$
$$m = 5$$

Now Try:

1.   Solve this equation and check the solution: $8n + 3 = 59$.

_____

Step 3 Check the solution by going back to the original equation. Replace m with 5.

$$7m + 2 = 37$$
$$7(5) + 2 = 37$$
$$35 + 2 = 37$$
$$37 = 37$$

When m is replaced with 5, the equation balances, so 5 is the correct solution.

2. Solve this equation and check the solution: $5x - 6 = 8x - 18$.

To keep the variable on the left side, add the opposite of $8x$, which is $-8x$ to both sides.

$$5x - 6 = 8x - 18$$
$$\underline{-8x -8x }$$
$$-3x - 6 = 0 - 18$$
$$-3x + (-6) = 0 + (-18)$$
$$\underline{ 6 6 }$$
$$-3x + 0 = 0 + (-12)$$
$$\frac{-3x}{-3} = \frac{-12}{-3}$$
$$x = 4$$

Check
$$5x - 6 = 8x - 18$$
$$5(4) - 6 = 8(4) - 18$$
$$20 + (-6) = 32 + (-18)$$
$$14 = 14$$

When x is replaced with 4, the equation balances, so 4 is the correct solution.

2. Solve this equation and check the solution: $8x - 15 = 10x + 5$.

_____

Review these examples for Objective 2:

3. Solve this equation and check the solution: $-12 = 6(y - 2)$.

We can use the distributive property to simplify the right side of the equation. Then use the steps to solve for y.

$$-12 = 6(y - 2)$$
$$-12 = 6 \cdot y - 6 \cdot 2$$
$$-12 = 6y - 12$$
$$-12 = 6y + (-12)$$
$$\underline{12 12 }$$
$$0 = 6y$$

Now Try:

3. Solve this equation and check the solution: $8(m - 3) = -24$.

_____

$$\frac{0}{6} = \frac{6y}{6}$$

$$0 = y$$

The solution is 0.

Check Go back to the original equation and replace y with 0.

$$-12 = 6(y - 2)$$

$$-12 = 6(0 - 2)$$

$$-12 = 6[0 + (-2)]$$

$$-12 = 6(-2)$$

$$-12 = -12$$

When y is replaced with 0, the equation balances, so 0 is the correct solution.

4. Solve this equation and check the solution: $3 + 6(n + 5) = 5 + 2n$.

Step 1 Use the distributive property on the left side.

$$3 + 6(n + 5) = 5 + 2n$$

$$3 + 6n + 30 = 5 + 2n$$

Step 2 Combine like terms on the left side.

$$6n + 33 = 5 + 2n$$

Step 3 Add $-2n$ to both sides.

$$\underline{-2n \qquad\quad -2n}$$

$$4n + 33 = 5 + 0$$

$$4n + 33 = 5$$

Step 3 To get $4n$ by itself, add -33 to both sides.

$$\underline{\quad -33 \quad -33}$$

$$4n + 0 = -28$$

Step 4 Divide both sides by 4, the coefficient of the variable term $4n$.

$$\frac{4n}{4} = \frac{-28}{4}$$

$$n = -7$$

Step 5 Check

$$3 + 6(n + 5) = 5 + 2n$$

$$3 + 6(-7 + 5) = 5 + 2(-7)$$

$$3 + 6(-2) = 5 + (-14)$$

$$3 + (-12) = -9$$

$$-9 = -9$$

When n is replaced with -7, the equation balances, so -7 is the correct solution.

4. Solve this equation and check the solution:

$$5 + 8(m + 2) = 5m + 6.$$

_____

Objective 1   Solve equations using the addition and division properties of equality.

For extra help, see Examples 1–2 on pages 142–143 of your text and Section Lecture video for Section 2.5 and Exercise Solutions Clip 1.

Solve each equation and then check.

1. $4m - 19 = -19$

1. _____

2. $-14 + 8d = -24 + 3d$

2. _____

3. $5q + 10 = 19 + 8q$

3. _____

Objective 2   Solve equations using the distributive, addition, and division properties.

For extra help, see Examples 3–4 on pages 144–145 of your text and Section Lecture video for Section 2.5 and Exercise Solutions Clip 25, 45, and 47.

Solve each equation and check.

4. $40 = 5(b + 8)$

4. _____

5. $t + 28 - 10 = 4(t + 6) - 24$

5. _____

6. $-87 + 3g = 9(g - 5) + 8g$

6. _____

Chapter 3 SOLVING APPLICATION PROBLEMS

3.1   Problem Solving: Perimeter

Learning Objectives
1   Use the formula for perimeter of a square to find the perimeter or the length of one side.
2   Use the formula for perimeter of a rectangle to find the perimeter, the length, or the width.
3   Find the perimeter of parallelograms, triangles, and irregular shapes.

Key Terms

Use the vocabulary terms listed below to complete each statement in exercises 1–6.

formula	**perimeter**	**square**
rectangle	**parallelogram**	**triangle**

1.   A figure with exactly three sides is a _____.

2.   The distance around the outside edges of a flat shape is the _____.

3.   A _____ is a rule for solving common types of problems.

4.   A figure with four sides that are all the same length and meet to form right angles is a _____.

5.   A four-sided figure in which all sides meet to form right angles is a

       _____.

6.   A four-sided figure in which opposite sides are both parallel and equal in length is

       a _____.

Guided Examples

Review these examples for Objective 1:

1.   Find the perimeter of the square that measures 7 ft on each side.

       Use the formula for perimeter of a square,
       $P = 4s$. Replace s with 7 ft.

       $P = 4s$

       $P = 4 \cdot 7$ ft

       $P = 28$ ft

       The perimeter of the square is 28 ft.

Now Try:

1.   Find the perimeter of the square that measures 15 in. on each side.

       _____

2. If the perimeter of a square is 120 ft, find the length of one side.

Use the formula for perimeter of a square, $P = 4s$. Replace P with 120 ft.

$$P = 4s$$

$$120 \text{ ft} = 4s$$

$$\frac{120 \text{ ft}}{4} = \frac{4s}{4}$$

$$30 \text{ ft} = s$$

The length of one side of the square is 30 ft.

2. If the perimeter of a square is 500 cm, find the length of one side.

_____

Review these examples for Objective 2:

3. Find the perimeter of a rectangle with length 32 m and width 17 m.

The replace l with 32 m and w with 17 m.

$$P = 2l + 2w$$

$$P = 2 \cdot 32 \text{ m} + 2 \cdot 17 \text{ m}$$

$$P = 64 \text{ m} + 34 \text{ m}$$

$$P = 98 \text{ m}$$

The perimeter of the rectangle is 98 m.

Check
To check the solution, add the lengths of the four sides.

$$P = 32 \text{ m} + 32 \text{ m} + 17 \text{ m} + 17 \text{ m}$$

$$P = 98 \text{ m}$$

Now Try:

3. Find the perimeter of a rectangle with length 54 ft and width 38 ft.

_____

4. If the perimeter of a rectangle is 32 ft and the width is 4 ft, find the length.

Use the formula $P = 2l + 2w$. Replace P with 32 ft and w with 4 ft.

$$P = 2l + 2w$$

$$32 \text{ ft} = 2l + 2 \cdot 4 \text{ ft}$$

$$32 \text{ ft} = 2l + 8 \text{ ft}$$

$$\underline{-8 \text{ ft} \qquad -8 \text{ ft}}$$

$$24 \text{ ft} = 2l$$

$$\frac{24 \text{ ft}}{2} = \frac{2l}{2}$$

$$12 \text{ ft} = l$$

The length is 12 ft.

4. If the perimeter of a rectangle is 48 in. and the width is 6 in., find the length.

_____

Name: Date:

Instructor: Section:

Review these examples for Objective 3:

5. Find the perimeter of the parallelogram with adjacent sides 15 m and 7 m.

To find the perimeter, add the lengths of the sides.

$$P = 15 \text{ m} + 15 \text{ m} + 7 \text{ m} + 7 \text{ m}$$
$$P = 44 \text{ m}$$

6. Find the perimeter of the triangle with sides 17 m, 8 m, and 15 m.

To find the perimeter, add the lengths of the sides.

$$P = 17 \text{ m} + 8 \text{ m} + 15 \text{ m}$$
$$P = 40 \text{ m}$$

7. Find the perimeter of the figure.

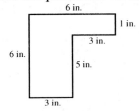

Find the perimeter by adding the lengths of the sides.

$$P = 6 \text{ in.} + 1 \text{ in.} + 3 \text{ in.} + 5 \text{ in.} + 3 \text{ in.} + 6 \text{ in.}$$
$$P = 24 \text{ in.}$$

Now Try:

5. Find the perimeter of the parallelogram with adjacent sides 25 m and 12 m.

_____

6. Find the perimeter of the triangle with sides 23 mm, 29 mm, and 18 mm.

_____

7. Find the perimeter of the figure.

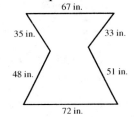

_____

Objective 1   **Use the formula for perimeter of a square to find the perimeter or the length of one side.**

For extra help, see Examples 1–2 on page 167 of your text and Section Lecture video for Section 3.1 and Exercise Solutions Clip 5.

Find the perimeter of each square, using the appropriate formula.

1.

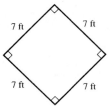

1. _____

2.

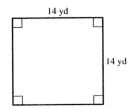

2. _____

For the given perimeter of the square, find the length of one side using the appropriate formula.

3. The perimeter is 76 miles.

3. _____

Objective 2 **Use the formula for perimeter of a rectangle to find the perimeter, the length, or the width.**

For extra help, see Examples 3–4 on pages 168–169 of your text and Section Lecture video for Section 3.1 and Exercise Solutions Clip 21, 27, and 29.

Find the perimeter of the rectangle, using the appropriate formula.

4.

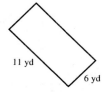

11 yd

6 yd

4. _____

For each rectangle, you are given the perimeter and either the length or the width. Find the unknown measurement by using the appropriate formula.

5. The perimeter is 90 ft and the length is 32 ft.

5. _____

6. The width is 17 cm and the perimeter is 76 cm.

6. _____

Objective 3 **Find the perimeter of parallelograms, triangles, and irregular shapes.**

For extra help, see Examples 5–7 on page 170 of your text and Section Lecture video for Section 3.1 and Exercise Solutions Clip 35 and 43.

Find the perimeter of each shape.

7. Parallelogram

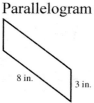

8 in.

3 in.

7. _____

8.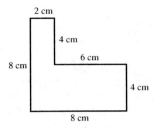

8. _____

9.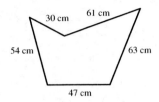

9. _____

Chapter 3 SOLVING APPLICATION PROBLEMS

3.2   Problem Solving: Area

Learning Objectives

1   Use the formula for area of a rectangle to find the area, the length, or the width.
2   Use the formula for area of a square to find the area or the length of one side.
3   Use the formula for area of a parallelogram to find the area, the base, or the height.
4   Solve application problems involving perimeter and area of rectangles, squares, or parallelograms.

Key Terms

Use the vocabulary terms listed below to complete each statement in exercises 1−4.

         area   **rectangle**   **square**   **parallelogram**

1.   To find the area of a_____, use the formula $A = s^2$.

2.   To find the area of a _____, use the formula $A = bh$.

3.   To find the area of a _____, use the formula $A = lw$.

4.   The surface inside a two-dimensional (flat) shape is the _____.

Guided Examples

Review these examples for Objective 1:	Now Try:
1. Find the area of each rectangle.	1. Find the area of each rectangle.
a. A rectangle measuring 9 m by 15 m.	**a.** A rectangle measuring 11 m by 23 m.
The length of this rectangle is 15 m and the width is 9 m. Use the formula $A = lw$ to find the area.	
$A = l \cdot w$	
$A = 15 \text{ m} \cdot 9 \text{ m}$	
$A = 135 \text{ m}^2$	
The area of the rectangle is 135 m^2.	

b. A rectangle measuring 14 cm by 8 cm.

The length of this rectangle is 14 cm and the width is 8 cm. Then use the formula $A = lw$.

$$A = l \cdot w$$

$$A = 14 \text{ cm} \cdot 8 \text{ cm}$$

$$A = 112 \text{ cm}^2$$

The area of the rectangle is 112 cm^2.

2. If the area of a rectangular rug is 24 yd^2, and the length is 6 yd, find the width.

Use the formula for area of a rectangle, $A = lw$.

Replace A with 24 yd^2, and replace l with 6 yd.

$$A = l \cdot w$$

$$24 \text{ yd}^2 = 6 \text{ yd} \cdot w$$

$$\frac{24 \text{ yd} \cdot \cancel{\text{yd}}}{6 \ \cancel{\text{yd}}} = \frac{6 \text{ yd} \cdot w}{6 \text{ yd}}$$

$$4 \text{ yd} = w$$

The width of the rug is 4 yd.

Check To check the solution, use the area formula. Replace l with 6 yd and replace w with 4 yd.

$$A = l \cdot w$$

$$A = 6 \text{ yd} \cdot 4 \text{ yd}$$

$$A = 24 \text{ yd}^2$$

An area of 24 yd^2 matches the information in the original problem. So 4 yd is the correct width of the rug.

Review this example for Objective 2:

3. Find the area of a square that is 5 ft on each side.

Use the formula for area of a square, $A = s^2$.
Replace s with 5 ft.

$$A = s^2$$

$$A = s \ \cdot \ s$$

$$A = 5 \text{ ft} \cdot 5 \text{ ft}$$

$$A = 25 \text{ ft}^2$$

The area of the square is 25 ft^2.

b. A rectangle measuring 27 cm by 7 cm.

_____

2. If the area of a rectangular flower garden is 35 ft^2, and the length is 7 ft, find the width.

_____

Now Try:

3. Find the area of a square that is 13 in. on each side.

_____

4. If the area of a square park is 81 mi^2, what is the length of one side of the park?

Use the formula for area of a square, $A = s^2$.
The value of A is 81 mi^2.

$$A = s^2$$

$$81 \text{ mi}^2 = s^2$$

$$81 \text{ mi}^2 = s \cdot s$$

$$81 \text{ mi}^2 = 9 \text{ mi} \cdot 9 \text{ mi}$$

The value of s is 9 mi, so the length of one side of the park is 9 mi.

4. If the area of a square park is 25 mi^2, what is the length of one side of the park?

_____

Review these examples for Objective 3:

5. Find the area of each parallelogram.

a. The base is 32 cm and the height is 15 cm.

Use the formula for the area of a parallelogram, $A = bh$. Replace b with 32 cm and replace h with 15 cm.

$$A = \quad b \quad \cdot \quad h$$

$$A = 32 \text{ cm} \cdot 15 \text{ cm}$$

$$A = 480 \text{ cm}^2$$

The area of the parallelogram is 480 cm^2.

b. The base is 53 m and the height is 18 m.

Use the formula for the area of a parallelogram, $A = bh$. Replace b with 53 m and replace h with 18 m.

$$A = \quad b \quad \cdot \quad h$$

$$A = 53 \text{ m} \cdot 18 \text{ m}$$

$$A = 954 \text{ m}^2$$

The area of the parallelogram is 954 m^2.

Now Try:

5. Find the area of each parallelogram.
a. The base is 19 cm and the height is 9 cm.

_____

b. The base is 25 m and height is 17 m.

_____

6. The area of a parallelogram is 48 ft^2 and the base is 8 ft. Find the height.

Use the formula for the area of a parallelogram, $A = bh$. The value of A is 48 ft^2, and the value of b is 8 ft.

$$A = b \cdot h$$

$$48 \text{ ft}^2 = 8 \text{ ft} \cdot h$$

$$\frac{48 \text{ ft} \cdot \cancel{ft}}{8 \; \cancel{ft}} = \frac{8 \text{ ft} \cdot h}{8 \text{ ft}}$$

$$6 \text{ ft} = h$$

The height of the parallelogram is 6 ft.

Check To check the solution use the area formula.

$$A = b \cdot h$$

$$A = 8 \text{ ft} \cdot 6 \text{ ft}$$

$$A = 48 \text{ ft}^2$$

An area of 48 ft^2 matches the information in the original problem. So 6 ft is the correct height of the parallelogram.

6. The area of a parallelogram is 54 ft^2 and the base is 6 ft. Find the height.

_____

Review this example for Objective 4:

7. Carla plans to adorn her afghan with a ribbon border. If the afghan measures 7 ft long and 5 ft wide, and the ribbon costs $3 per linear foot, how much is the total cost for this decoration?

The ribbon will go around the edges of the afghan, so you need to find the perimeter of the afghan. Replace *l* with 7 ft and replace *w* with 5 ft.

$$P = 2l + 2w$$

$$P = 2 \cdot 7 \text{ ft} + 2 \cdot 5 \text{ ft}$$

$$P = 14 \text{ ft} + 10 \text{ ft}$$

$$P = 24 \text{ ft}$$

The perimeter of the afghan is 24 ft, so Carla will need 24 ft of ribbon. The cost of the ribbon is $3 per linear foot, which means $3 for 1 foot. To find the cost for 24 ft, multiply $3 \cdot 24 \text{ ft}$. The ribbon will cost $72.

Now Try:

7. A soybean field is 332 m wide and 512 m long. Find the perimeter and area of the field.

_____

Name: Date:

Instructor: Section:

Objective 1   Use the formula for area of a rectangle to find the area, the length, or the width.

For extra help, see Examples 1–2 on pages 176–177 of your text and Section Lecture video for Section 3.2 and Exercise Solutions Clip 21.

Find the area of each rectangle using the appropriate formula.

1. 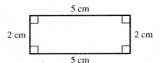 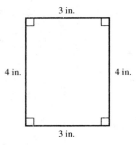   1. _____

2.   2. _____

Use the area of the rectangle and either its length or width, and the appropriate formula, to find the other measurement.

3.   The area of a food tray is 4104 cm^2 and the width is   3. _____
      54 cm. Find its length.

Objective 2   Use the formula for area of a square to find the area or the length of one side.

For extra help, see Examples 3–4 on pages 177–178 of your text and Section Lecture video for Section 3.2 and Exercise Solutions Clip 7.

Find the area of each square using the appropriate formula.

4.   4. _____

5.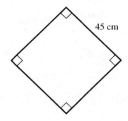

5. _____

Given the area of the square, find the length of one side by inspection.

6. The area of a square pot holder is 49 in.^{2}.

6. _____

Objective 3   **Use the formula for area of a parallelogram to find the area, the base, or the height.**

For extra help, see Examples 5–6 on pages 179–180 of your text and Section Lecture video for Section 3.2 and Exercise Solutions Clip 9, 13, and 27.

Find the area of each parallelogram using the appropriate formula.

7.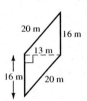

7. _____

8.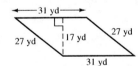

8. _____

Use the area of the parallelogram and either its base or height, and the appropriate formula, to find the other measurement.

9. The area is 72 m^2, and the height is 12 m. Find the base.

9. _____

Objective 4   **Solve application problems involving perimeter and area of rectangles, squares, or parallelograms.**

For extra help, see Example 7 on page 180 of your text and Section Lecture video for Section 3.2 and Exercise Solutions Clip 39.

Solve each application problem. You may need to find the perimeter, the area, or one of the side measurements

10.   Chuck and Carla plan to carpet their 8 m by 4 m living room. If carpeting costs $9 per square meter, how much will carpeting this room cost?

10. _____

11.   Tabitha plans to sew a quilt measuring 88 ft^2 in area. If the quilt must be 11 ft long, how wide will the quilt be?

11. _____

12.   A rectangular dance floor measures 27 yd long and 15 yd wide. To cover the floor in wood paneling, it would cost $12 per square yard. How much is the total cost for covering this dance floor in wood paneling?

12. _____

Chapter 3 SOLVING APPLICATION PROBLEMS

3.3   Solving Application Problems with One Unknown Quantity

Learning Objectives
1   Translate word phrases into algebraic expressions.
2   Translate sentences into equations.
3   Solve application problems with one unknown quantity.

Key Terms

Use the vocabulary terms listed below to complete each statement in exercises 1−4.

sum	difference	product	quotient
increased by	less than	double	per

1. _____ and _____ are words that mean addition.

2. _____ and _____ are words that mean multiplication.

3. _____ and _____ are words that mean division.

4. _____ and _____ are words that mean subtraction.

Guided Examples

Review these examples for Objective 1:

1. Write each phrase as an algebraic expression. Use x as the variable.

   a. A number plus 15

   Algebraic expression: $x + 15$ or $15 + x$

   b. The sum of 18 and a number

   Algebraic expression: $18 + x$ or $x + 18$

   c. 110 more than a number

   Algebraic expression: $x + 110$ or $110 + x$

   d. −66 added to a number

   Algebraic expression: $-66 + x$ or $x + (-66)$

Now Try:

1. Write each phrase as an algebraic expression. Use x as the variable.

   a. A number plus 7

   _____

   b. The sum of 43 and a number

   _____

   c. 63 more than a number

   _____

   d. −20 added to a number

   _____

e. A number increased by 39

Algebraic expression: $x + 39$ or $39 + x$

f. 17 less than a number

Algebraic expression: $x - 17$

g. A number subtracted from 73

Algebraic expression: $73 - x$

h. 73 subtracted from a number

Algebraic expression: $x - 73$

i. 34 fewer than a number

Algebraic expression: $x - 34$

j. A number decreased by 51

Algebraic expression: $x - 51$

k. 39 minus a number

Algebraic expression: $39 - x$

2. Write each phrase as an algebraic expression. Use x as the variable.

a. 112 times a number

Algebraic expression: $112x$

b. The product of 25 and a number

Algebraic expression: $25x$

c. Triple a number

Algebraic expression: $3x$

d. The quotient of -7 and a number

Algebraic expression: $\dfrac{-7}{x}$

e. A number divided by 9

Algebraic expression: $\dfrac{x}{9}$

e. A number increased by 75

_____

f. 44 less than a number

_____

g. A number subtracted from 89

_____

h. 89 subtracted from a number

_____

i. 16 fewer than a number

_____

j. A number decreased by 36

_____

k. 48 minus a number

_____

2. Write each phrase as an algebraic expression. Use x as the variable.

a. 57 times a number

_____

b. The product of 43 and a number

_____

c. Double a number

_____

d. The quotient of -13 and a number

_____

e. A number divided by 19

_____

f. 17 subtracted from 8 times a number

Algebraic expression: $8x - 17$

f. 81 subtracted from 6 times a number

_____

Review this example for Objective 2:

3. If 6 times a number is added to 13, the result is 37. Find the number. Let x represent the number.

Let x represent the unknown number. Use the information in the problem to write an equation.

6 times a number added to 13 is 37.
$$6x \quad + \quad 13 = 37$$

Next, solve the equation.

$$6x + 13 = 37$$
$$\underline{-13 \qquad -13}$$
$$6x + 0 = 24$$
$$\frac{6x}{6} = \frac{24}{6}$$
$$x = 4$$

The number is 4.

Check   Go back to the words of the original problem.

6 times a number added to 13 is 37.
$$6 \cdot \quad 4 \quad + \quad 13 = 37$$

Since $6 \cdot 4 + 13 = 24 + 13 = 37,$ then 4 is the correct solution.

Now Try:

3. If 7 times a number is added to 21, the result is 56. Write the equation and find the number.

_____

Review these examples for Objective 3:

4. Shawn was paid a $100 bonus check at work, which he promptly deposited into his checking account. After buying $47 worth of groceries and concert tickets that cost a total of $64, he found that he had a total of $219 in his account. How much was in his account before he deposited the check?

Step 1 Read the problem. Find the balance of the bank account.

Step 2 Assign a variable. There is only one unknown quantity. Let a represent the account balance.

Now Try:

4. Betty left 9 pounds of bird seed in her garage one evening. During the night, mice ate some of the bird seed. Later, Betty bought a 30-pound bag of bird seed which she added to the bird seed left in the garage. At this point, she had 36 pounds of bird seed. How much bird seed did the mice eat?

_____

Step 3 Write an equation.

Account balance	put in $100	took out $47	took out $64	Ended up with $219
a	$+100$	-47	-64	$=219$

Step 4 Solve.

$$a+100 \ - \ 47 \ - \ 64 = 219$$
$$a+100+(-47)+(-64) = 219$$
$$a+ \ \ 53 \ \ \ +(-64) = 219$$
$$a \ \ \ \ \ -11 \ \ \ \ = 219$$
$$\underline{ +11 \ \ \ \ \ \ +11 }$$
$$a \ + \ 0 \ \ \ \ = 230$$
$$a = \ 230$$

Step 5 State the answer. The original account balance was $230.

Step 6 Check the solution. Go back to the original problem and insert the solution.
Start with $230 in the account.
Put in $100, so $230 + $100 = $330.
Took out $47, so $330 – $47 = $283.
Took out $64, so $283 – $64 = $219.
Ended with $219. This checks.

5. An office supply store bought nine boxes of black ink printer cartridges. One day the store sold 13 cartridges and the next day the store sold 22 cartridges. If nineteen cartridges were left on the shelf, how many cartridges were in each box?

Step 1 Read the problem. Find the number of cartridges in each box.

Step 2 Assign a variable. There is only one unknown quantity. Let *n* represent the number of cartridges in each box.

Step 3 Write an equation.

Number of boxes	cartridges in each box	took out 13	took out 22	Ended with 19
9	$\cdot$ $\quad n$	-13	-22	$=19$

5. Before a camping weekend, Angela bought 4 packages of paper plates. On Saturday, one package was used up while on Sunday 30 paper plates were used. If two packages plus 25 plates remain, how many paper plates were there in each package?

_____

Step 4 Solve.

$$9n \quad - \quad 13 \quad - \quad 22 = \quad 19$$
$$9n + (-13) + (-22) = \quad 19$$
$$9n + \quad \quad (-35) \quad \quad = \quad 19$$
$$\underline{\quad \quad \quad +35 \quad \quad \quad \quad +35}$$
$$9n \quad + \quad 0 \quad \quad \quad = \quad 54$$
$$\frac{9n}{9} \quad \quad \quad = \quad \frac{54}{9}$$
$$n = \quad 6$$

Step 5 State the answer. Each box contains 6 cartridges.

Step 6 Check the solution.
9 boxes each contains 6 cartridges, so $9 \cdot 6 = 54$.
13 were sold, so $54 - 13 = 41$.
Then 22 were sold, so $41 - 22 = 19$.
Ended with 19. This checks.

6. While grocery shopping, Teresa spent $9 less than four times what Pepe spent. If Teresa spent $23, how much did Pepe spend?

Step 1 Read the problem. Find the amount Pepe spent.

Step 2 Assign a variable. There is only one unknown quantity. Let x represent the amount Pepe spent.

Step 3 Write an equation.
$9 less than four times what Pepe spent, is $23
$$4x - 9 = 23$$

Step 4 Solve.
$$4x - 9 = 23$$
$$\underline{\quad +9 \quad +9}$$
$$4x + 0 = 32$$
$$\frac{4x}{4} = \frac{32}{4}$$
$$x = 8$$

Step 5 State the answer. Pepe spent $8.

Step 6 Check the solution. Go back to the original problem and insert the solution.
$$4 \cdot 8 - 9 = 32 - 9 = 23$$
This checks.

6. The price of a DVD is $3.00 less than twice the cost of a book. If the DVD costs $25.00, how much does the book cost?

_____

Objective 1   Translate word phrases into algebraic expressions.

For extra help, see Examples 1–2 on page 187 of your text and Section Lecture video for Section 3.3 and Exercise Solutions Clip 5, 7, and 19.

Write an algebraic expression using x as the variable.

1.   The product of –6 and a number   1. _____

2.   The quotient of a number and 10   2. _____

3.   One more than three times a number   3. _____

Objective 2   Translate sentences into equations.

For extra help, see Example 3 on page 188 of your text and Section Lecture video for Section 3.3 and Exercise Solutions Clip 25 and 27.

Translate each sentence into an equation and solve it. Check your solution by going back to the words in the original problem.

4.   If three times a number is decreased by eight, the   4.
      result is 58. Find the number.   Equation _____

                                                                     Solution _____

5.   The sum of three and seven times a number is 31.   5.
      Find the number.   Equation _____

                                                                     Solution _____

6.   If seven times a number is subtracted from twelve   6.
      times the number, the result is –30. Find the number.   Equation _____

                                                                     Solution _____

Name: Date:

Instructor: Section:

Objective 3   Solve application problems with one unknown quantity.

For extra help, see Examples 4–6 on pages 189–191 of your text and Section Lecture video for Section 3.3 and Exercise Solutions Clip 35.

Solve each application problem. Use the six problem-solving steps listed in your text.

7. The number of bananas an adult gorilla ate is 15 less than three times the number a younger gorilla ate. If the adult gorilla ate 21 bananas, how many did the younger gorilla eat?

7. _____

8. A local jeweler sold three gold watches for the same retail price and established a checking account with this money. The jeweler then bought a diamond for $4000 and a pearl necklace for $11,000 with money from this account. The jeweler now realized this new account was overdrawn for $6000. For how much did she sell each of these gold watches?

8. _____

9. There were 12 celery sticks in the vegetable crisper in Lance's refrigerator. After eating some of these celery sticks, Lance added 16 carrot sticks to the crisper. Later, Lance noticed there were a total of 21 celery and carrot sticks altogether. How many celery sticks did Lance eat?

9. _____

Chapter 3 SOLVING APPLICATION PROBLEMS

3.4   Solving Application Problems with Two Unknown Quantities

Learning Objectives
1   Solve application problems with two unknown quantities.

Key Terms

Use the vocabulary terms listed below to complete each statement in exercises 1−4.

added to	subtracted from	times	divided by
minus	more than	half	triple

1. _____ and _____ are words that mean addition.

2. _____ and _____ are words that mean multiplication.

3. _____ and _____ are words that mean division.

4. _____ and _____ are words that mean subtraction.

Guided Examples

Review these examples for Objective 1:

1. Last year Lucas earned $5000 less than his wife, Susanna. Together they earned $75,320. How much did each of them earn?

    Step 1 Read the problem. Find the amount Lucas earned and the amount Susanna earned.

    Step 2 Assign a variable. There are two unknowns. Let s represent the amount Susanna earned. Let $s - 5000$ represent the amount Lucas earned.

    Step 3 Write an equation.

Susanna's earnings		Lucas' earning		Total earnings
s	$+$	$s - 5000$	$=$	$75,320$

Now Try:

1. Linda paid three times as much for her dog than for her cat. If she paid a total of $220 for both, how much did she pay for each pet?

    _____

Step 4 Solve.

$$s + s - 5000 = 75{,}320$$
$$2s + (-5000) = 75{,}320$$
$$\underline{+5000 +5000}$$
$$2s + 0 = 80{,}320$$
$$\frac{2s}{2} = \frac{80{,}320}{2}$$
$$s = 40{,}160$$

Step 5 State the answer. Susanna earned $40,160.

$s - 5000$ represents Lucas' earnings. Replace s with $40,160.

$40,160 – $5000 = $35,160, so Lucas earned $35,160.

Step 6 Check the solution. The total earnings is $40, 160 + $35,160 = $75,320. This checks.

2. A string is 89 cm long. Bob's cat, Reginald, bit the string into two pieces so that one piece is 17 cm longer than the other. Find the length of each piece.

Step 1 Read the problem. Find the length of a longer piece and a shorter piece.

Step 2 Assign a variable. There are two unknowns. Let x represent the length of the shorter piece. Let $x + 17$ represent the length of the longer piece.

Step 3 Write an equation.

shorter piece		longer piece		Total length
x	$+$	$x + 17$	$=$	89

Step 4 Solve.

$$x + x + 17 = 89$$
$$2x + 17 = 89$$
$$\underline{-17 -17}$$
$$2x + 0 = 72$$
$$\frac{2x}{2} = \frac{72}{2}$$
$$x = 36$$

2. A telephone cable 98 meters in length is cut into two pieces. If one piece is 22 meters longer than the other, how long are the two pieces?

_____

Step 5 State the answer. The shorter piece is 36 cm.

$x + 17$ represents the longer piece. Replace x with 36.

$36 + 17 = 53$, so the longer piece is 53 cm.

Step 6 Check the solution. The total length is 36 cm + 53 cm = 89 cm. This checks.

3. A site of a new building is rectangular in shape. The length is twice the width. If it will require 642 meters of fence to enclose the site, find the length and the width of the new building site.

    Step 1 Read the problem. Find the length and the width.

    Step 2 Assign a variable. There are two unknowns. Let x represent the width. Let $2x$ represent the length.

    Step 3 Write an equation. Use the formula for perimeter of a rectangle, $P = 2l + 2w$. Replace P with 642, replace l with $2x$, and replace w with x.

    $$P = 2\ l\ + 2w$$
    $$642 = 2(2x) + 2x$$

    Step 4 Solve.
    $$642 = 2(2x) + 2x$$
    $$642 = 4x + 2x$$
    $$\frac{642}{6} = \frac{6x}{6}$$
    $$107 = x$$

    Step 5 State the answer. The width is 107 m. $2x$ represents the length. Replace x with 107. $2 \cdot 107 = 214$, so the length is 214 m.

    Step 6 Check the solution.
    $$P = 2 \cdot 214 + 2 \cdot 107$$
    $$P = \ \ 428\ +\ 214$$
    $$P = \ \ \ \ \ \ \ \ 642$$
    This matches the perimeter given in the original problem.

3. The length of a picture frame is 5 in. less than three times the width. If the perimeter of the frame is 86 in., find the length and the width of the picture frame.

    _____

Name: Date:

Instructor: Section:

Objective 1   Solve application problems with two unknown quantities.

For extra help, see Examples 1–3 on pages 196–199 of your text and Section Lecture video for Section 3.4 and Exercise Solutions Clip 5, 13, and 21.

Solve each application problem by using the six problem-solving steps.

1.   The vote totals for the two candidates for county judge, Jones and Greer, differed by only 516 votes, with Jones receiving more votes. If there were 16,786 voters in the county, how many votes did each candidate receive?

      1. Greer_____

           Jones_____

2.   After a 14-meter tall tree is chopped down, it is cut into four pieces. Three pieces are the same length, while the fourth piece is 2 m longer than each of the other three. Find the length of each piece.

      2. _____

Use the formula $P = 2l + 2w$ to solve. Make a sketch to help you solve each problem.

3.   A rectangular garden is three times as long as it is wide. The perimeter of the garden is 96 yd. Find the length and the width of the garden.

      3. width_____

           length _____

Chapter 4 RATIONAL NUMBERS: POSITIVE AND NEGATIVE FRACTIONS

4.1   Introduction to Signed Fractions

Learning Objectives
1   Use a fraction to name part of a whole.
2   Identify numerators, denominators, proper fractions, and improper fractions.
3   Graph positive and negative fractions on a number line.
4   Find the absolute value of a fraction.
5   Write equivalent fractions.

Key Terms

Use the vocabulary terms listed below to complete each statement in exercises 1–6.

fraction	numerator	denominator
proper fraction	**improper fraction**	**equivalent fractions**

1. Two fractions are _____ when they represent the same portion of a whole.

2. A number of the form $\frac{a}{b}$, where a and b are integers and $b \neq 0$ is called a

    _____.

3. A fraction whose numerator is larger than its denominator is called an

    _____.

4. In the fraction $\frac{2}{9}$, the 2 is the _____.

5. A fraction whose denominator is larger than its numerator is called a

    _____.

6. The _____ of a fraction shows the number of equal parts in a whole.

Name: Date:
Instructor: Section:

Guided Examples

Review these examples for Objective 1:

1. Use fractions to represent the shaded portion and the unshaded portion of the figure.

    The figure has 8 equal parts. The 3 shaded parts are represented by the fraction $\frac{3}{8}$. The unshaded part is $\frac{5}{8}$ of the figure.

2. Use a fraction to represent the shaded parts.

    An area equal to 5 of the $\frac{1}{2}$ parts is shaded, so $\frac{5}{2}$ is shaded.

Now Try:

1. Use fractions to represent the shaded portion and the unshaded portion of the figure.

    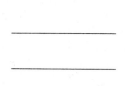

    _____

    _____

2. Use a fraction to represent the shaded parts.

    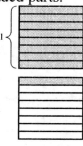

    _____

Review these examples for Objective 2:

3. Identify the numerator and denominator in each fraction. Then state the number of equal parts in the whole.

    a. $\frac{9}{14}$

        $\underline{9}$ ← Numerator
        14 ← Denominator
        14 equal parts in the whole

    b. $\frac{4}{15}$

        $\underline{4}$ ← Numerator
        15 ← Denominator
        15 equal parts in the whole

Now Try:

3. Identify the numerator and denominator in each fraction. Then state the number of equal parts in the whole.

    a. $\frac{5}{17}$

        _____

        _____

    b. $\frac{6}{23}$

        _____

        _____

4. $\dfrac{4}{7}, \dfrac{7}{10}, \dfrac{19}{6}, \dfrac{11}{11}, \dfrac{13}{27}, \dfrac{1}{8}, \dfrac{6}{5}$

   a. Identify all the proper fractions in the list.

   Proper fractions have a numerator that is less than the denominator.

   $\dfrac{4}{7}, \dfrac{7}{10}, \dfrac{13}{27}, \dfrac{1}{8}$

   b. Identify all the improper fractions in the list.

   Improper fractions have a numerator that is equal or greater than the denominator.

   $\dfrac{19}{6}, \dfrac{11}{11}, \dfrac{6}{5}$

4. $\dfrac{7}{8}, \dfrac{9}{11}, \dfrac{22}{7}, \dfrac{13}{13}, \dfrac{12}{29}, \dfrac{1}{5}, \dfrac{17}{4}$

   a. Identify all the proper fractions in the list.

   _____

   b. Identify all the improper fractions in the list.

   _____

Review this example for Objective 3:

5. Graph the fractions on the number line.

   $-\dfrac{7}{9}, \dfrac{7}{9}$

   For $-\dfrac{7}{9}$, the fraction is negative, so it is between 0 and −1. We divide that space into 9 equal parts. Then we start at 0 and count to the left 7 parts.

   There is no sign in front of $\dfrac{7}{9}$, so it is positive.

   Because $\dfrac{7}{9}$, is between 0 and 1, we divide that space into 9 equal parts. Then we start at 0 and count to the right 7 parts.

   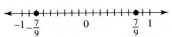

Now Try:

5. Graph the fractions on the number line.

   $\dfrac{5}{6}, -\dfrac{5}{6}$

   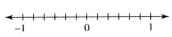

Review this example for Objective 4:

6. Find each absolute value: $\left| -\dfrac{3}{5} \right|$ and $\left| \dfrac{3}{5} \right|$.

   The distance from 0 to $-\dfrac{3}{5}$ on the number line is $\dfrac{3}{5}$ space, so $\left| -\dfrac{3}{5} \right| = \dfrac{3}{5}$.

   The distance from 0 to $\dfrac{3}{5}$ on the number line is also $\dfrac{3}{5}$ space, so $\left| \dfrac{3}{5} \right| = \dfrac{3}{5}$.

Now Try:

6. Find each absolute value: $\left| \dfrac{9}{10} \right|$ and $\left| -\dfrac{9}{10} \right|$.

   _____

   _____

Review these examples for Objective 5:

7.

a. Write $-\dfrac{1}{3}$ as an equivalent fraction with a denominator of 18.

In other words, $-\dfrac{1}{3} = -\dfrac{?}{18}$.

The original denominator is 3. Multiplying 3 times 6 gives 18, the new denominator. To write an equivalent fraction, multiply both the numerator and denominator by 6.

$$-\dfrac{1}{3} = -\dfrac{1 \cdot 6}{3 \cdot 6} = -\dfrac{6}{18}$$

So, $-\dfrac{1}{3}$ is equivalent to $-\dfrac{6}{18}$.

b. Write $\dfrac{15}{18}$ as an equivalent fraction with a denominator of 6.

In other words, $\dfrac{15}{18} = \dfrac{?}{6}$.

The original denominator is 18. Dividing 18 by 3 gives 6, the new denominator. To write an equivalent fraction, divide both the numerator and denominator by 3.

$$\dfrac{15}{18} = \dfrac{15 \div 3}{18 \div 3} = \dfrac{5}{6}$$

So, $\dfrac{15}{18}$ is equivalent to $\dfrac{5}{6}$.

8. Simplify each fraction by dividing the numerator by the denominator.

a. $\dfrac{12}{12}$

Think of $\dfrac{12}{12}$ as $12 \div 12$. The result is 1,

so $\dfrac{12}{12} = 1$.

b. $-\dfrac{20}{5}$

Think of $-\dfrac{20}{5}$ as $-20 \div 5$. The result is -4,

so $-\dfrac{20}{5} = -4$.

Now Try:

7.

a. Write $-\dfrac{4}{5}$ as an equivalent fraction with a denominator of 30.

_____

b. Write $\dfrac{35}{42}$ as an equivalent fraction with a denominator of 6.

_____

8. Simplify each fraction by dividing the numerator by the denominator.

a. $\dfrac{22}{22}$

_____

b. $-\dfrac{42}{7}$

_____

c. $\dfrac{7}{1}$ **c.** $\dfrac{13}{1}$

Think of $\dfrac{7}{1}$ as $7 \div 1$. The result is 7,

so $\dfrac{7}{1} = 7$.   _____

Objective 1   Use a fraction to name part of a whole.

For extra help, see Examples 1–2 on pages 218–219 of your text and Section Lecture video for Section 4.1 and Exercise Solutions Clip 3 and 5.

Write the fractions that represent the shaded and unshaded portions of each figure.

1.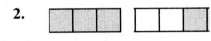

1. Shaded _____

   Unshaded _____

2.

2. Shaded _____

   Unshaded _____

3.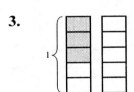

3. Shaded _____

   Unshaded _____

Objective 2   Identify numerators, denominators, proper fractions, and improper fractions.

For extra help, see Examples 3–4 on pages 219–220 of your text and Section Lecture video for Section 4.1 and Exercise Solutions Clip 15 and 19.

Identify the numerator and denominator in each fraction.

4. $\dfrac{2}{9}$

4. numerator _____

   denominator _____

5. $\dfrac{12}{5}$

5. numerator _____

   denominator _____

List the proper and improper fractions in the group of numbers.

6. $\dfrac{2}{3}, \dfrac{5}{2}, \dfrac{4}{4}, \dfrac{6}{7}, \dfrac{8}{15}, \dfrac{20}{19}$

6. **Proper** _____

   Improper _____

Objective 3   Graph positive and negative fractions on a number line.

For extra help, see Example 5 on page 221 of your text and Section Lecture video for Section 4.1 and Exercise Solutions Clip 21.

Graph each pair of fractions on a number line.

7. $\dfrac{2}{3}, -\dfrac{2}{3}$

7.

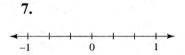

8. $-\dfrac{1}{4}, \dfrac{1}{4}$

8.

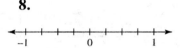

9. $-\dfrac{5}{8}, \dfrac{5}{8}$

9.

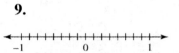

Objective 4   Find the absolute value of a fraction.

For extra help, see Example 6 on page 221 of your text and Section Lecture video for Section 4.1 and Exercise Solutions Clip 31 and 34.

Find each absolute value.

10. $\left| -\dfrac{15}{7} \right|$

10. _____

11. $\left| \dfrac{8}{9} \right|$

11. _____

12. $\left| 0 \right|$

12. _____

Name: Date:
Instructor: Section:

Objective 5   Write equivalent fractions.

For extra help, see Examples 7–8 on pages 223–224 of your text and Section Lecture video for Section 4.1 and Exercise Solutions Clip 35 and 37.

Rewrite each fraction as an equivalent fraction with a denominator of 48.

13. $-\dfrac{2}{3}$ 13. _____

14. $\dfrac{3}{4}$ 14. _____

Rewrite the fraction as an equivalent fraction with a denominator of 5.

15. $-\dfrac{4}{20}$ 15. _____

Chapter 4 RATIONAL NUMBERS: POSITIVE AND NEGATIVE FRACTIONS

4.2 Writing Fractions in Lowest Terms

Learning Objectives
1   Identify fractions written in lowest terms.
2   Write a fraction in lowest terms using common factors.
3   Write a number as a product of prime factors.
4   Write a fraction in lowest terms using prime factorization.
5   Write a fraction with variables in lowest terms.

Key Terms

Use the vocabulary terms listed below to complete each statement in exercises 1–4.

     lowest terms   **prime number**   **composite number**

     prime factorization

1.   A _____ has at least one factor other than itself and 1.

2.   In a _____ every factor is a prime number.

3.   The factors of a _____ are itself and 1.

4.   A fraction is written in _____ when its numerator and denominator have no common factor other than 1.

Guided Examples

Review these examples for Objective 1:

1. Are the following fractions in lowest terms? If not, find a common factor of the numerator and denominator (other than 1).

    a. $\dfrac{3}{7}$

    The numerator and denominator have no common factor other than 1, so the fraction is in lowest terms.

    b. $\dfrac{28}{49}$

    The numerator and denominator have a common factor of 7, so the fraction is not in lowest terms.

Now Try:

1. Are the following fractions in lowest terms? If not, find a common factor of the numerator and denominator (other than 1).

    a. $\dfrac{8}{9}$

    _____

    b. $\dfrac{27}{36}$

    _____

Review these examples for Objective 2:

2. Divide by a common factor to write each fraction in lowest terms.

    a. $\dfrac{40}{48}$

    The greatest common factor of 40 and 48 is 8. Divide both numerator and denominator by 8.
    $$\frac{40}{48} = \frac{40 \div 8}{48 \div 8} = \frac{5}{6}$$

    b. $\dfrac{35}{55}$

    Divide both numerator and denominator by 5.
    $$\frac{35}{55} = \frac{35 \div 5}{55 \div 5} = \frac{7}{11}$$

    c. $-\dfrac{28}{49}$

    Divide both numerator and denominator by 7. Keep the negative sign.
    $$-\frac{28}{49} = -\frac{28 \div 7}{49 \div 7} = -\frac{4}{7}$$

    d. $\dfrac{80}{96}$

    Dividing by 4 does not give the greatest common factor of 80 and 96.
    $$\frac{80}{96} = \frac{80 \div 4}{96 \div 4} = \frac{20}{24} \quad \leftarrow \text{not lowest terms}$$
    There is another common factor of 4.
    $$\frac{20}{24} = \frac{20 \div 4}{24 \div 4} = \frac{5}{6} \quad \leftarrow \text{lowest terms}$$
    Divide by 16, the greatest common factor, in one step.
    $$\frac{80}{96} = \frac{80 \div 16}{96 \div 16} = \frac{5}{6}$$

Now Try:

2. Divide by a common factor to write each fraction in lowest terms.

    a. $\dfrac{20}{36}$

    _____

    b. $\dfrac{35}{40}$

    _____

    c. $-\dfrac{25}{65}$

    _____

    d. $\dfrac{55}{60}$

    _____

Review these examples for Objective 3:

3. Label each number as prime or composite or neither.

    1 3 4 6 17 23

    First, 1 is neither prime nor composite.
    Next, 3, 17, and 23 are prime.
    The numbers 4 and 6 can be divided by 2, so 4 and 6 are composite.

Now Try:

3. Label each number as prime or composite or neither.

    1 8 21 29 31

    _____

4. Find the prime factorization of each number.

a. 54

Start by dividing 54 by 2 and work your way up the chain of divisions.

$$
\begin{array}{r}
1 \\
3\overline{)3} \\
3\overline{)9} \\
3\overline{)27} \\
\text{Start here} \quad 2\overline{)54}
\end{array}
$$

Because all the factors (divisors) are prime, the prime factorization of 54 is $54 = 2 \cdot 3 \cdot 3 \cdot 3$.

b. 84

Start by dividing 84 by 2.

$$
\begin{array}{r}
1 \\
7\overline{)7} \\
3\overline{)21} \\
2\overline{)42} \\
\text{Start here} \quad 2\overline{)84}
\end{array}
$$

The prime factorization of 84 is $84 = 2 \cdot 2 \cdot 3 \cdot 7$.

5. Find the prime factorization of each number.

a. 180

Try to divide 180 by 2. The quotient is 90. Write the factors 2 and 90 under 180. Box the 2 (or circle it), because it is prime.

$$
\begin{array}{c}
180 \\
\swarrow \quad \searrow \\
\boxed{2} \quad\quad 90
\end{array}
$$

Try dividing 90 by 2. The quotient is 45. Write the factors 2 and 45 under the 90. Continue until you find the prime factorization.

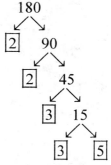

The prime factorization is $180 = 2 \cdot 2 \cdot 3 \cdot 3 \cdot 5$.

4. Find the prime factorization of each number.

a. 96

_____

b. 56

_____

5. Find the prime factorization of each number.

a. 152

_____

b. 108   **b.** 168

Divide 108 by 2, the first prime number.

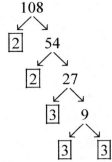

The prime factorization is $108 = 2 \cdot 2 \cdot 3 \cdot 3 \cdot 3$.

Review these examples for Objective 4:

6. Write each fraction in lowest terms.

a. $\dfrac{28}{63}$

Write the prime factorizations of 28 and 63.

$$\frac{28}{63} = \frac{2 \cdot 2 \cdot 7}{3 \cdot 3 \cdot 7}$$

Divide the numerator and denominator by 7, the common factor.

$$\frac{28}{63} = \frac{2 \cdot 2 \cdot \overset{1}{\cancel{7}}}{3 \cdot 3 \cdot \underset{1}{\cancel{7}}} = \frac{4}{9}$$

b. $\dfrac{90}{126}$

Write the prime factorizations of 90 and 126. Then divide the numerator and denominator by the common factors.

$$\frac{90}{126} = \frac{\overset{1}{\cancel{2}} \cdot \overset{1}{\cancel{3}} \cdot \overset{1}{\cancel{3}} \cdot 5}{\underset{1}{\cancel{2}} \cdot \underset{1}{\cancel{3}} \cdot \underset{1}{\cancel{3}} \cdot 7} = \frac{5}{7}$$

c. $\dfrac{27}{180}$

$$\frac{27}{180} = \frac{3 \cdot \overset{1}{\cancel{3}} \cdot \overset{1}{\cancel{3}}}{2 \cdot 2 \cdot \underset{1}{\cancel{3}} \cdot \underset{1}{\cancel{3}} \cdot 5} = \frac{3}{20}$$

Now Try:

6. Write each fraction in lowest terms.

a. $\dfrac{75}{225}$

_____

b. $\dfrac{42}{140}$

_____

c. $\dfrac{36}{45}$

_____

Name: Date:
Instructor: Section:

Review these examples for Objective 5:

7. Write each fraction in lowest terms.

   a. $\dfrac{10}{15x}$

$$\frac{10}{15x} = \frac{2 \cdot \cancel{5}^{\,1}}{3 \cdot \cancel{5} \cdot x_{\,1}} = \frac{2}{3x}$$

   b. $\dfrac{7xy}{28xy}$

$$\frac{7xy}{28xy} = \frac{\cancel{7}^{\,1} \cdot \cancel{x}^{\,1} \cdot \cancel{y}^{\,1}}{2 \cdot 2 \cdot \cancel{7} \cdot \cancel{x} \cdot \cancel{y}_{\,1\;\;1\;\;\;1}} = \frac{1}{4}$$

   c. $\dfrac{6b^4}{12ab^2}$

$$\frac{6b^4}{12ab^2} = \frac{\cancel{2}^{\,1} \cdot \cancel{3}^{\,1} \cdot b \cdot b \cdot \cancel{b}^{\,1} \cdot \cancel{b}^{\,1}}{\cancel{2} \cdot 2 \cdot \cancel{3} \cdot a \cdot \cancel{b} \cdot \cancel{b}} = \frac{b^2}{2a}$$

   d. $\dfrac{5a^2b}{x^2}$

$$\frac{5a^2b}{x^2} = \frac{5 \cdot a \cdot a \cdot b}{x \cdot x}$$

$\dfrac{5a^2b}{x^2}$ is already in lowest terms.

Now Try:

7. Write each fraction in lowest terms.

   a. $\dfrac{25}{30x}$

_____

   b. $\dfrac{8ab}{24ab}$

_____

   c. $\dfrac{9b^5}{36ab^3}$

_____

   d. $\dfrac{9p^2q}{r^3}$

_____

Objective 1   Identify fractions written in lowest terms.

For extra help, see Example 1 on page 229 of your text and Section Lecture video for Section 4.2.

Are the following fractions in lowest terms? If not, find a common factor of the numerator and denominator (other than 1).

1. $-\dfrac{7}{11}$ **1.** _____

2. $\dfrac{20}{25}$ **2.** _____

3. $\dfrac{12}{16}$ **3.** _____

Name: Date:
Instructor: Section:

Objective 2   Write a fraction in lowest terms using common factors.

For extra help, see Example 2 on page 230 of your text and Section Lecture video for Section 4.2 and Exercise Solutions Clip 3.

Write in lowest terms.

4. $-\dfrac{24}{36}$ 4. _____

5. $\dfrac{30}{42}$ 5. _____

6. $-\dfrac{32}{72}$ 6. _____

Objective 3   Write a number as a product of prime factors.

For extra help, see Examples 3–5 on pages 231–233 of your text and Section Lecture video for Section 4.2 and Exercise Solutions Clip 5 and 13.

Find the prime factorization of each number.

7. 105   7. _____

8. 168   8. _____

9. 320   9. _____

Objective 4   Write a fraction in lowest terms using prime factorization.

For extra help, see Example 6 on pages 234–235 of your text and Section Lecture video for Section 4.2.

Write each numerator and denominator as a product of prime factors. Then use the prime factorization to write the fraction in lowest terms.

10. $\dfrac{18}{99}$ 10. _____

11. $\dfrac{63}{105}$ 11. _____

12. $\dfrac{36}{210}$ 12. _____

Objective 5   Write a fraction with variables in lowest terms.

For extra help, see Example 7 on page 236 of your text and Section Lecture video for Section 4.2 and Exercise Solutions Clip 49, 61, and 65.

Write each fraction in lowest terms.

13. $\dfrac{12r^2s}{4rs^3}$ 13. _____

14. $\dfrac{16b^2cd}{40b^2d}$ 14. _____

15. $\dfrac{8xy^2}{6x^2y^2}$ 15. _____

Chapter 4 RATIONAL NUMBERS: POSITIVE AND NEGATIVE FRACTIONS

4.3   Multiplying and Dividing Signed Fractions

Learning Objectives
1   Multiply signed fractions.
2   Multiply fractions that involve variables.
3   Divide signed fractions.
4   Divide fractions that involve variables.
5   Solve application problems involving multiplying and dividing fractions.

Key Terms

Use the vocabulary terms listed below to complete each statement in exercises 1−4.

reciprocals   indicator words   of   each

1. Two numbers are _____ of each other if their product is 1.

2. The words "per" and "divided equally" are _____ for division.

3. When the word "_____" follows a fraction, it means "multiply".

4. The word "_____" indicates division.

Guided Examples

Review these examples for Objective 1:

1. Find each product.

a. $-\dfrac{7}{9} \cdot -\dfrac{5}{11}$

Multiply the numerators and multiply the denominators.

$$-\frac{7}{9} \cdot -\frac{5}{11} = \frac{7 \cdot 5}{9 \cdot 11} = \frac{35}{99}$$

The answer is in lowest terms because 35 and 99 have no common factor other than 1.

b. $\left(\dfrac{3}{5}\right)\left(-\dfrac{4}{7}\right)$

$$\left(\frac{3}{5}\right)\left(-\frac{4}{7}\right) = -\frac{3 \cdot 4}{5 \cdot 7} = -\frac{12}{35}$$

Now Try:

1. Find each product.

a. $-\dfrac{10}{11} \cdot -\dfrac{4}{13}$

_____

b. $\left(\dfrac{15}{22}\right)\left(-\dfrac{3}{4}\right)$

_____

2. Multiply. Write the products in lowest terms.

a. $-\dfrac{9}{7}\left(\dfrac{7}{15}\right)$

Multiplying a negative number times a positive number gives a negative product.

$$-\dfrac{9}{7}\left(\dfrac{7}{15}\right) = -\dfrac{3\cdot 3\cdot 7}{7\cdot 3\cdot 5} = -\dfrac{\overset{1}{\cancel{3}}\cdot 3\cdot \overset{1}{\cancel{7}}}{\cancel{7}\cdot \cancel{3}\cdot 5} = -\dfrac{3}{5}$$

b. Find $\dfrac{3}{8}$ of $\dfrac{4}{9}$.

Recall that "of" indicates multiplication.

$$\dfrac{3}{8}\cdot\dfrac{4}{9} = \dfrac{3\cdot 2\cdot 2}{2\cdot 2\cdot 2\cdot 3\cdot 3} = \dfrac{\overset{1}{\cancel{3}}\cdot\overset{1}{\cancel{2}}\cdot\overset{1}{\cancel{2}}}{\cancel{2}\cdot\cancel{2}\cdot 2\cdot\cancel{3}\cdot 3} = \dfrac{1}{6}$$

3. Find $\dfrac{3}{7}$ of 14.

Write 14 in fraction form as $\dfrac{14}{1}$.

$$\dfrac{3}{7}\cdot\dfrac{14}{1} = \dfrac{3\cdot 2\cdot 7}{7\cdot 1} = -\dfrac{3\cdot 2\cdot\overset{1}{\cancel{7}}}{\cancel{7}\cdot 1} = \dfrac{6}{1} = 6$$

Review these examples for Objective 2:

4. Find each product.

a. $\dfrac{6x}{7}\cdot\dfrac{5}{18x}$

$$\dfrac{6x}{7}\cdot\dfrac{5}{18x} = \dfrac{2\cdot 3\cdot x\cdot 5}{7\cdot 2\cdot 3\cdot 3\cdot x} = \dfrac{\overset{1}{\cancel{2}}\cdot\overset{1}{\cancel{3}}\cdot\overset{1}{\cancel{x}}\cdot 5}{7\cdot\cancel{2}\cdot\cancel{3}\cdot 3\cdot\cancel{x}} = \dfrac{5}{21}$$

b. $\left(\dfrac{9a}{5b}\right)\left(\dfrac{10b^2}{3a}\right)$

$$\left(\dfrac{9a}{5b}\right)\left(\dfrac{10b^2}{3a}\right) = \dfrac{3\cdot 3\cdot a\cdot 2\cdot 5\cdot b\cdot b}{5\cdot b\cdot 3\cdot a}$$

$$= \dfrac{\overset{1}{\cancel{3}}\cdot 3\cdot\overset{1}{\cancel{5}}\cdot 2\cdot\overset{1}{\cancel{b}}\cdot\overset{1}{\cancel{b}}\cdot b}{\cancel{5}\cdot\cancel{b}\cdot\cancel{3}\cdot\cancel{a}}$$

$$= 6b$$

2. Multiply. Write the products in lowest terms.

a. $-\dfrac{11}{7}\left(\dfrac{14}{33}\right)$

_____

b. Find $\dfrac{2}{7}$ of $\dfrac{21}{40}$.

_____

3. Find $\dfrac{5}{9}$ of 27.

_____

Now Try:

4. Find each product.

a. $\dfrac{12x}{5}\cdot\dfrac{7}{18x}$

_____

b. $\left(\dfrac{7x}{9c}\right)\left(\dfrac{6c^2}{35x}\right)$

_____

Name: _____   Date: _____
Instructor: _____   Section: _____

Review these examples for Objective 3:

5. Rewrite each division problem as a multiplication problem.

a. $\dfrac{5}{7} \div \dfrac{10}{3}$

$$\dfrac{5}{7} \div \dfrac{10}{3} = \dfrac{5}{7} \cdot \dfrac{3}{10} = \dfrac{\overset{1}{\cancel{5}} \cdot 3}{7 \cdot 2 \cdot \underset{1}{\cancel{5}}} = \dfrac{3}{14}$$

b. $6 \div \left(-\dfrac{1}{5}\right)$

$$6 \div \left(-\dfrac{1}{5}\right) = \dfrac{6}{1} \cdot \left(-\dfrac{5}{1}\right) = -\dfrac{6 \cdot 5}{1 \cdot 1} = -\dfrac{30}{1} = -30$$

c. $-\dfrac{7}{9} \div (-6)$

$$-\dfrac{7}{9} \div (-6) = -\dfrac{7}{9} \cdot \left(-\dfrac{1}{6}\right) = \dfrac{7 \cdot 1}{3 \cdot 3 \cdot 3 \cdot 2} = \dfrac{7}{54}$$

d. $\dfrac{5}{11} \div 0$

$\dfrac{5}{11} \div 0$ is undefined.

e. $0 \div \dfrac{7}{11}$

$$0 \div \dfrac{7}{11} = 0 \cdot \dfrac{11}{7} = 0$$

Review these examples for Objective 4:

6. Divide. (Assume that none of the variables represent zero.)

a. $\dfrac{x^3}{y^2} \div \dfrac{x^2}{4y}$

$$\dfrac{x^3}{y^2} \div \dfrac{x^2}{4y} = \dfrac{x^3}{y^2} \cdot \dfrac{4y}{x^2} = \dfrac{\overset{1}{\cancel{x}} \cdot \overset{1}{\cancel{x}} \cdot x \cdot 2 \cdot 2 \cdot \overset{1}{\cancel{y}}}{\underset{1}{\cancel{y}} \cdot y \cdot \underset{1}{\cancel{x}} \cdot \underset{1}{\cancel{x}}} = \dfrac{4x}{y}$$

b. $\dfrac{9b^2}{7} \div b^3$

$$\dfrac{9b^2}{7} \div b^3 = \dfrac{9b^2}{7} \cdot \dfrac{1}{b^3} = \dfrac{3 \cdot 3 \cdot \overset{1}{\cancel{b}} \cdot \overset{1}{\cancel{b}} \cdot 1}{7 \cdot \underset{1}{\cancel{b}} \cdot \underset{1}{\cancel{b}} \cdot b} = \dfrac{9}{7b}$$

Now Try:

5. Rewrite each division problem as a multiplication problem.

a. $\dfrac{5}{9} \div \dfrac{20}{3}$

_____

b. $10 \div \left(-\dfrac{1}{8}\right)$

_____

c. $-\dfrac{6}{13} \div \left(-\dfrac{3}{26}\right)$

_____

d. $\dfrac{13}{27} \div 0$

_____

e. $0 \div \dfrac{2}{25}$

_____

Now Try:

6. Divide. (Assume that none of the variables represent zero.)

a. $\dfrac{a^2}{b^3} \div \dfrac{a}{5b^2}$

_____

b. $\dfrac{10b}{9} \div b^3$

_____

Review these examples for Objective 5:

7. Solve each application problem.

    a. Signe gives $\frac{1}{12}$ of her income to charities. Last year she earned \$48,000. How much did she give to charities?

    Because the word "of" follows the fraction, it indicates multiplication.

    $$\frac{1}{12} \cdot 48,000 = \frac{1}{12} \cdot \frac{48,000}{1} = \frac{1 \cdot \cancel{12} \cdot 4000}{\cancel{12} \cdot 1} = 4000$$

    She gave \$4000 to charities.

    b. How many times can a $1\frac{2}{3}$-quart spray bottle be filled before 20 quarts of water are used up?

    $$20 \div 1\frac{2}{3} = 20 \div \frac{5}{3} = \frac{20}{1} \cdot \frac{3}{5} = \frac{4 \cdot \cancel{5} \cdot 3}{1 \cdot \cancel{5}} = 12$$

    The spray bottle can be filled 12 times.

Now Try:

7. Solve each application problem.

    a. A flower bed is $\frac{12}{11}$ m by $\frac{7}{8}$ m. Find its area.

    _____

    b. How many $\frac{1}{3}$-lb servings will a 24-lb turkey provide?

    _____

Objective 1   Multiply signed fractions.

For extra help, see Examples 1–3 on pages 241–244 of your text and Section Lecture video for Section 4.3 and Exercise Solutions Clip 3, 5, and 7.

Multiply. Write the products in lowest terms.

1. $-\frac{10}{42} \cdot \frac{3}{5}$

    1. _____

2. $\frac{6}{18} \cdot \frac{9}{2}$

    2. _____

3. $\frac{5}{9}$ of 81

    3. _____

Name: Date:

Instructor: Section:

Objective 2   Multiply fractions that involve variables.

For extra help, see Example 4 on page 244 of your text and Section Lecture video for Section 4.3 and Exercise Solutions Clip 13.

Use prime factorization to find these products.

4. $\dfrac{3c}{5} \cdot \dfrac{c}{9}$ 4. _____

5. $\left(\dfrac{m}{10}\right)\left(\dfrac{25}{m^2}\right)$ 5. _____

6. $\dfrac{5a}{7b^2} \cdot \dfrac{14b}{10a^2}$ 6. _____

Objective 3   Divide signed fractions.

For extra help, see Example 5 on pages 246–247 of your text and Section Lecture video for Section 4.3 and Exercise Solutions Clip 15, 19, and 31.

Divide. Write the quotients in lowest terms.

7. $\dfrac{7}{8} \div (-21)$ 7. _____

8. $-\dfrac{5}{12} \div \dfrac{15}{8}$ 8. _____

9. $-\dfrac{2}{3} \div \left(-\dfrac{7}{9}\right)$ 9. _____

Objective 4   Divide fractions that involve variables.

For extra help, see Example 6 on page 247 of your text and Section Lecture video for Section 4.3 and Exercise Solutions Clip 23 and 25.

Divide. Write the quotients in lowest terms.

10. $\dfrac{3b}{5a} \div \dfrac{6}{7ab}$ 10. _____

11. $10x^2 \div \dfrac{5x}{3}$ 11. _____

12. $\dfrac{7c}{3d} \div 14c^2 d$ 12. _____

Objective 5   Solve application problems involving multiplying and dividing fractions.

For extra help, see Example 7 on page 248 of your text and Section Lecture video for Section 4.3.

Solve each application problem.

13.   Phillip saves $\frac{2}{9}$ of his income each month. How   13. _____
        much did he save last month if he earned $2700?

14.   Rochelle sells hats at craft shows. She needs $\frac{2}{5}$ yd   14.
        for each hat. How many hats can she make from 10
        yards of fabric?

Chapter 4 RATIONAL NUMBERS: POSITIVE AND NEGATIVE FRACTIONS

4.4   Adding and Subtracting Signed Fractions

Learning Objectives
1   Add and subtract like fractions.
2   Find the lowest common denominator for unlike fractions.
3   Add and subtract unlike fractions.
4   Add and subtract unlike fractions that contain variables.

Key Terms

Use the vocabulary terms listed below to complete each statement in exercises 1−3.

like fractions   unlike fractions   least common denominator

1.   Fractions with different denominators are called _____.

2.   Fractions with the same denominator are called _____.

3.   The _____ of two whole numbers is the smallest whole number divisible by both of the numbers.

Guided Examples

Review these examples for Objective 1:

1.   Find each sum or difference.

   a. $\dfrac{1}{9} + \dfrac{5}{9}$

   These are like fractions because they have a common denominator. Add the numerators and write the sum over the common denominator.

   $$\dfrac{1}{9} + \dfrac{5}{9} = \dfrac{1+5}{9} = \dfrac{6}{9}$$

   Now write $\dfrac{6}{9}$ in lowest terms.

   $$\dfrac{6}{9} = \dfrac{2 \cdot \cancel{3}^{\,1}}{3 \cdot \cancel{3}_{\,1}} = \dfrac{2}{3}$$

   b. $-\dfrac{4}{7} + \dfrac{6}{7}$

   $$-\dfrac{4}{7} + \dfrac{6}{7} = \dfrac{-4+6}{7} = \dfrac{2}{7}$$

Now Try:

1.   Find each sum or difference.

   a. $\dfrac{1}{12} + \dfrac{7}{12}$

   _____

   b. $-\dfrac{1}{4} + \dfrac{3}{4}$

   _____

c. $\dfrac{4}{15} - \dfrac{7}{15}$

Write the subtraction as adding the opposite.

$$\dfrac{4}{15} - \dfrac{7}{15} = \dfrac{4-7}{15} = \dfrac{4+(-7)}{15} = \dfrac{-3}{15}, \text{ or } -\dfrac{3}{15}$$

Now write $-\dfrac{3}{15}$ in lowest terms.

$$-\dfrac{3}{15} = -\dfrac{\overset{1}{\cancel{3}}}{\underset{1}{\cancel{3} \cdot 5}} = -\dfrac{1}{5}$$

d. $\dfrac{6}{x^3} - \dfrac{4}{x^3}$

$$\dfrac{6}{x^3} - \dfrac{4}{x^3} = \dfrac{6-4}{x^3} = \dfrac{2}{x^3}$$

---

c. $\dfrac{7}{20} - \dfrac{19}{20}$

_____

d. $\dfrac{9}{a^2} - \dfrac{5}{a^2}$

_____

Review these examples for Objective 2:

2.

a. Find the LCD for $\dfrac{2}{7}$ and $\dfrac{5}{21}$ by inspection.

Since 21 is divisible by 7, then 21 is the LCD.

b. Find the LCD for $\dfrac{7}{12}$ and $\dfrac{3}{8}$ by inspection.

Since 12 is not divisible by 8, use multiples of 12, that is 12, 24, and 36. Notice that 24 is divisible by both 8 and 12. The LCD is 24.

3.

a. What is the LCD for $\dfrac{11}{18}$ and $\dfrac{5}{24}$?

Write 18 and 24 as the product of prime factors.
$$18 = 2 \cdot 3 \cdot 3$$
$$24 = 2 \cdot 2 \cdot 2 \cdot 3$$
$$\text{LCD} = 2 \cdot 2 \cdot 2 \cdot 3 \cdot 3 = 72$$

The LCD for $\dfrac{11}{18}$ and $\dfrac{5}{24}$ is 72.

b. What is the LCD for $\dfrac{9}{16}$ and $\dfrac{7}{36}$?

$$16 = 2 \cdot 2 \cdot 2 \cdot 2$$
$$36 = 2 \cdot 2 \cdot 3 \cdot 3$$
$$\text{LCD} = 2 \cdot 2 \cdot 2 \cdot 2 \cdot 3 \cdot 3 = 144$$

The LCD for $\dfrac{9}{16}$ and $\dfrac{7}{36}$ is 144.

Now Try:

2.

a. Find the LCD for $\dfrac{2}{9}$ and $\dfrac{7}{27}$ by inspection.

_____

b. Find the LCD for $\dfrac{3}{10}$ and $\dfrac{7}{15}$ by inspection.

_____

3.

a. What is the LCD for $\dfrac{9}{20}$ and $\dfrac{8}{25}$?

_____

b. What is the LCD for $\dfrac{7}{30}$ and $\dfrac{2}{45}$?

_____

Review these examples for Objective 3:

4. Find each sum or difference.

    a. $\dfrac{1}{2}+\dfrac{1}{6}$

Step 1 The larger denominator (6) is the LCD.

Step 2 $\dfrac{1}{2}=\dfrac{1\cdot 3}{2\cdot 3}=\dfrac{3}{6}$ and $\dfrac{1}{6}$ already has the LCD.

Step 3 Add the numerators.

$$\frac{1}{2}+\frac{1}{6}=\frac{3}{6}+\frac{1}{6}=\frac{3+1}{6}=\frac{4}{6}$$

Step 4 Write $\dfrac{4}{6}$ in lowest terms.

$$\frac{4}{6}=\frac{2\cdot \cancel{2}^{\,1}}{\cancel{2}_{\,1}\cdot 3}=\frac{2}{3}$$

    b. $\dfrac{3}{8}-\dfrac{7}{12}$

Step 1 The LCD is 24.

Step 2 $\dfrac{3}{8}=\dfrac{3\cdot 3}{8\cdot 3}=\dfrac{9}{24}$ and $\dfrac{7}{12}=\dfrac{7\cdot 2}{12\cdot 2}=\dfrac{14}{24}$

Step 3 Subtract the numerators.

$$\frac{3}{8}-\frac{7}{12}=\frac{9}{24}-\frac{14}{24}=\frac{9-14}{24}=\frac{-5}{24},\text{ or } -\frac{5}{24}$$

Step 4 $-\dfrac{5}{24}$ is in lowest terms.

    c. $-\dfrac{7}{18}+\dfrac{5}{12}$

Step 1 Use prime factorization to find the LCD.

$$18=2\cdot 3\cdot 3$$
$$12=2\cdot 2\cdot 3$$
$$\text{LCD}=2\cdot 2\cdot 3\cdot 3=36$$

Step 2

$$-\frac{7}{18}=-\frac{7\cdot 2}{18\cdot 2}=-\frac{14}{36}\text{ and }\frac{5}{12}=\frac{5\cdot 3}{12\cdot 3}=\frac{15}{36}$$

Step 3 Add the numerators.

$$-\frac{7}{18}+\frac{5}{12}=-\frac{14}{36}+\frac{15}{36}=\frac{-14+15}{36}=\frac{1}{36}$$

Step 4 $\dfrac{1}{36}$ is in lowest terms.

Now Try:

4. Find each sum or difference.

    a. $\dfrac{5}{9}+\dfrac{7}{18}$

_____

    b. $\dfrac{8}{15}-\dfrac{7}{10}$

_____

    c. $-\dfrac{5}{24}+\dfrac{7}{9}$

_____

d. $5 - \dfrac{3}{4}$

Step 1 The LCD for $\dfrac{5}{1}$ and $\dfrac{3}{4}$ is 4, the larger denominator.

Step 2 $\dfrac{5}{1} = \dfrac{5 \cdot 4}{1 \cdot 4} = \dfrac{20}{4}$ and $\dfrac{3}{4}$ already has the LCD.

Step 3 Subtract the numerators.

$$\dfrac{5}{1} - \dfrac{3}{4} = \dfrac{20}{4} - \dfrac{3}{4} = \dfrac{20-3}{4} = \dfrac{17}{4}$$

Step 4 $\dfrac{17}{4}$ is in lowest terms.

d. $9 - \dfrac{2}{7}$

_____

Review these examples for Objective 4:

5. Find each sum or difference.

   a. $\dfrac{1}{3} + \dfrac{x}{2}$

Step 1 The LCD is 6.

Step 2 $\dfrac{1}{3} = \dfrac{1 \cdot 2}{3 \cdot 2} = \dfrac{2}{6}$ and $\dfrac{x}{2} = \dfrac{x \cdot 3}{2 \cdot 3} = \dfrac{3x}{6}$

Step 3 $\dfrac{1}{3} + \dfrac{x}{2} = \dfrac{2}{6} + \dfrac{3x}{6} = \dfrac{2+3x}{6}$

Step 4 $\dfrac{2+3x}{6}$ is in lowest terms.

   b. $\dfrac{5}{6} - \dfrac{7}{x}$

Step 1 The LCD is $6x$.

Step 2 $\dfrac{5}{6} = \dfrac{5 \cdot x}{6 \cdot x} = \dfrac{5x}{6x}$ and $\dfrac{7}{x} = \dfrac{7 \cdot 6}{x \cdot 6} = \dfrac{42}{6x}$

Step 3 $\dfrac{5}{6} - \dfrac{7}{x} = \dfrac{5x}{6x} - \dfrac{42}{6x} = \dfrac{5x-42}{6x}$

Step 4 $\dfrac{5x-42}{6x}$ is in lowest terms.

Now Try:

5. Find each sum or difference.

   a. $\dfrac{2}{3} + \dfrac{x}{5}$

_____

   b. $\dfrac{5}{8} - \dfrac{8}{x}$

_____

Objective 1   Add and subtract like fractions.

For extra help, see Example 1 on page 254 of your text and Section Lecture video for Section 4.4 and Exercise Solutions Clip 13.

Write each sum or difference in lowest terms.

1. $-\dfrac{14}{15} + \dfrac{4}{15}$

1. _____

2. $\dfrac{3}{4a} + \dfrac{1}{4a}$ 2. _____

3. $\dfrac{7}{y^2} - \dfrac{3}{y^2}$ 3. _____

Objective 2   Find the lowest common denominator for unlike fractions.

For extra help, see Examples 2–3 on page 256 of your text and Section Lecture video for Section 4.4.

Find the LCD for each pair of fractions.

4. $\dfrac{3}{7}$ and $\dfrac{3}{14}$ 4. _____

5. $\dfrac{8}{21}$ and $\dfrac{8}{9}$ 5. _____

6. $\dfrac{7}{12}$ and $\dfrac{3}{40}$ 6. _____

Objective 3   Add and subtract unlike fractions.

For extra help, see Example 4 on pages 257–258 of your text and Section Lecture video for Section 4.4 and Exercise Solutions Clip 11, 17, and 19.

Find each sum or difference. Write all answers in lowest terms.

7. $\dfrac{1}{6} + \dfrac{2}{15}$ 7. _____

8. $-\dfrac{1}{2} + \dfrac{7}{12}$ 8. _____

9. $\dfrac{33}{40} - \dfrac{7}{24}$ 9. _____

Objective 4   Add and subtract unlike fractions that contain variables.

For extra help, see Example 5 on pages 258–259 of your text and Section Lecture video for Section 4.4 and Exercise Solutions Clip 21 and 23.

Find each sum or difference. Write all answers in lowest terms.

10. $\dfrac{3}{n}+\dfrac{3}{5}$ 10. _____

11. $\dfrac{1}{6}+\dfrac{x}{3}$ 11. _____

12. $\dfrac{3}{a}-\dfrac{b}{4}$ 12. _____

Chapter 4 RATIONAL NUMBERS: POSITIVE AND NEGATIVE FRACTIONS

4.5 Problem Solving: Mixed Numbers and Estimating

Learning Objectives
1   Identify mixed numbers and graph them on a number line.
2   Rewrite mixed numbers as improper fractions, or the reverse.
3   Estimate the answer and multiply or divide mixed numbers.
4   Estimate the answer and add or subtract mixed numbers.
5   Solve application problems containing mixed numbers.

Key Terms

Use the vocabulary terms listed below to complete each statement in exercises 1–2.

mixed number   **improper fraction**

1. A(n) _____ includes a fraction and a whole number written together.

2. A mixed number can be rewritten as a(n) _____.

Guided Examples

Review this example for Objective 1:

1. Use a number line to show mixed numbers $\frac{5}{2}$ and $-\frac{5}{2}$.

$\frac{5}{2}$ is equivalent to $2\frac{1}{2}$.

$-\frac{5}{2}$ is equivalent to $-2\frac{1}{2}$.

Now Try:

1. Use a number line to show mixed numbers $1\frac{1}{4}$ and $-1\frac{1}{4}$.

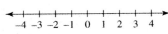

Review these examples for Objective 2:

2. Write $9\frac{3}{4}$ as an improper fraction.

Step 1 $9\frac{3}{4}$ $4 \cdot 9 = 36$ Then $36 + 3 = 39$

Step 2 $9\frac{3}{4} = \frac{39}{4}$

Now Try:

2. Write $5\frac{5}{6}$ as an improper fraction.

_____

3. Write each improper fraction as an equivalent mixed number in simplest form.

a. $\dfrac{18}{7}$

Divide 18 by 7.

$$7\overline{)18} \\ \quad\underline{14} \\ \quad\ 4$$ with quotient 2

The quotient 2 is the whole number part. The remainder 4 is the numerator of the fraction, and the denominator stays as 7.

$$\frac{18}{7} = 2\frac{4}{7}$$

b. $\dfrac{22}{6}$

Divide 22 by 6.

$$6\overline{)22} \quad \text{so} \quad \frac{22}{6} = 3\frac{4}{6} = 3\frac{2}{3} \\ \ \underline{18} \\ \ \ 4$$ with quotient 3

Or, write $\dfrac{22}{6}$ in lowest terms first.

$$\frac{22}{6} = \frac{\overset{1}{\cancel{2}} \cdot 11}{\underset{1}{\cancel{2}} \cdot 3} = \frac{11}{3} \quad \text{Then } 3\overline{)11} \text{ so } \frac{11}{3} = 3\frac{2}{3} \\ \qquad\qquad\qquad\qquad\quad \underline{9} \\ \qquad\qquad\qquad\qquad\quad 2$$ with quotient 3

3. Write each improper fraction as an equivalent mixed number in simplest form.

a. $\dfrac{27}{8}$

_____

b. $\dfrac{37}{9}$

_____

Review these examples for Objective 3:

4. Round each mixed number to the nearest whole number.

a. $1\dfrac{7}{12}$

$$1\frac{7}{12} \begin{array}{l} \leftarrow 7 \text{ is more than } 6 \\ \leftarrow \text{Half of 12 is 6} \end{array}$$

$1\dfrac{7}{12}$ rounds up to 2.

b. $8\dfrac{3}{7}$

$$8\frac{3}{7} \begin{array}{l} \leftarrow 3 \text{ is less than } 3\frac{1}{2} \\ \leftarrow \text{Half of 7 is } 3\frac{1}{2} \end{array}$$

$8\dfrac{3}{7}$ rounds down to 8.

Now Try:

4. Round each mixed number to the nearest whole number.

a. $6\dfrac{11}{20}$

_____

b. $4\dfrac{13}{28}$

_____

5. First, round the numbers and estimate each answer. Then find the exact answer. Write exact answers in simplest form.

a. $2\frac{1}{3} \cdot 4\frac{5}{7}$

Estimate the answer by rounding the mixed numbers.

$2\frac{1}{3}$ rounds to 2 and $4\frac{5}{7}$ rounds to 5

$2 \cdot 5 = 10 \leftarrow$ Estimated answer

To find the exact answer, first rewrite each mixed number as an improper fraction.

$2\frac{1}{3} = \frac{7}{3}$ and $4\frac{5}{7} = \frac{33}{7}$

Next, multiply.

$$2\frac{1}{3} \cdot 4\frac{5}{7} = \frac{7}{3} \cdot \frac{33}{7} = \frac{\overset{1}{\cancel{7}} \cdot \overset{1}{\cancel{3}} \cdot 11}{\underset{1}{\cancel{3}} \cdot \underset{1}{\cancel{7}}} = \frac{11}{1} = 11$$

The estimate was 10, so an exact answer of 11 is reasonable.

b. $5\frac{2}{5} \cdot 1\frac{1}{9}$

First, round each mixed number and estimate the answer.

$5\frac{2}{5}$ rounds to 5 and $1\frac{1}{9}$ rounds to 1

$5 \cdot 1 = 5 \leftarrow$ Estimated answer

Now find the exact answer.

$5\frac{2}{5} = \frac{27}{5}$ and $1\frac{1}{9} = \frac{10}{9}$

$$5\frac{2}{5} \cdot 1\frac{1}{9} = \frac{27}{5} \cdot \frac{10}{9} = \frac{\overset{1}{\cancel{9}} \cdot 3 \cdot 2 \cdot \overset{1}{\cancel{5}}}{\underset{1}{\cancel{5}} \cdot \underset{1}{\cancel{9}}} = \frac{6}{1} = 6$$

The estimate was 5, so an exact answer of 6 is reasonable.

5. First, round the numbers and estimate each answer. Then find the exact answer. Write exact answers in simplest form.

a. $4\frac{9}{10} \cdot 3\frac{4}{7}$

Estimate _____

Exact _____

b. $6\frac{1}{8} \cdot 5\frac{5}{7}$

Estimate _____

Exact _____

6. First, round the numbers and estimate each answer. Then find the exact answer. Write exact answers in simplest form.

a. $3\dfrac{1}{5} \div 2\dfrac{2}{3}$

Estimate the answer by rounding the mixed numbers.

$3\dfrac{1}{5}$ rounds to 3   and $2\dfrac{2}{3}$ rounds to 3

$3 \div 3 = 1 \leftarrow$ Estimated answer

To find the exact answer, first rewrite each mixed number as an improper fraction.

$$3\dfrac{1}{5} \div 2\dfrac{2}{3} = \dfrac{16}{5} \div \dfrac{8}{3}$$

Now rewrite the problem as multiplying by the reciprocal of $\dfrac{8}{3}$.

$$\dfrac{16}{5} \div \dfrac{8}{3} = \dfrac{16}{5} \cdot \dfrac{3}{8} = \dfrac{\overset{1}{\cancel{8}} \cdot 2 \cdot 3}{5 \cdot \underset{1}{\cancel{8}}} = \dfrac{6}{5} = 1\dfrac{1}{5}$$

The estimate was 1, so an exact answer of $1\dfrac{1}{5}$ is reasonable.

b. $7\dfrac{5}{7} \div 9$

First, round the numbers and estimate the answer.

$7\dfrac{5}{7}$ rounds to 8

$8 \div 9 = \dfrac{8}{9} \leftarrow$ Estimated answer

Now find the exact answer.

$$7\dfrac{5}{7} \div 9 = \dfrac{54}{7} \div \dfrac{9}{1} = \dfrac{54}{7} \cdot \dfrac{1}{9} = \dfrac{\overset{1}{\cancel{9}} \cdot 6}{7 \cdot \underset{1}{\cancel{9}}} = \dfrac{6}{7}$$

The estimate was $\dfrac{8}{9}$, so an exact answer of $\dfrac{6}{7}$ is reasonable.

6. First, round the numbers and estimate each answer. Then find the exact answer. Write exact answers in simplest form.

a. $4\dfrac{4}{9} \div 6\dfrac{2}{3}$

Estimate _____

Exact _____

b. $6\dfrac{7}{8} \div 11$

Estimate _____

Exact _____

Name: Date:

Instructor: Section:

Review these examples for Objective 4:

7. First, estimate each answer. Then add or subtract to find the exact answer. Write exact answers in simplest form.

    a. $3\frac{3}{10} + 4\frac{4}{5}$

To estimate the answer, round each mixed number to the nearest whole number.

$$3\frac{3}{10} + 4\frac{4}{5}$$

$$3 \;+\; 5 = 8 \;\leftarrow \text{Estimate}$$

To find the exact answer, first rewrite each mixed number as an equivalent improper fraction.

$$3\frac{3}{10} + 4\frac{4}{5} = \frac{33}{10} + \frac{24}{5}$$

The LCD for $\frac{33}{10}$ and $\frac{24}{5}$ is 10. Rewrite $\frac{24}{5}$ as an equivalent fraction with a denominator of 10.

$$\frac{33}{10} + \frac{24}{5} = \frac{33}{10} + \frac{48}{10} = \frac{33+48}{10} = \frac{81}{10} = 8\frac{1}{10}$$

The estimate was 8, so an exact answer of $8\frac{1}{10}$ is reasonable.

    b. $5\frac{1}{4} - 3\frac{5}{6}$

Round each number and estimate the answer.

$$5\frac{1}{4} - 3\frac{5}{6}$$

$$5 \;-\; 4 = 1 \;\leftarrow \text{Estimate}$$

To find the exact answer, rewrite the mixed numbers as improper fractions and subtract.

$$5\frac{1}{4} - 3\frac{5}{6} = \frac{21}{4} - \frac{23}{6} = \frac{63}{12} - \frac{46}{12}$$

$$= \frac{63-46}{12} = \frac{17}{12} = 1\frac{5}{12}$$

The estimate was 1, so an exact answer of $1\frac{5}{12}$ is reasonable.

Now Try:

7. First, estimate each answer. Then add or subtract to find the exact answer. Write exact answers in simplest form.

    a. $1\frac{2}{9} + 5\frac{1}{3}$

    Estimate   _____

    Exact   _____

    b. $12\frac{5}{12} - 8\frac{5}{6}$

    Estimate   _____

    Exact   _____

c. $7 - 3\frac{2}{5}$

$7 - 3\frac{2}{5}$

$7 - 3 = 4 \leftarrow$ Estimate

$7 - 3\frac{2}{5} = \frac{7}{1} - \frac{17}{5} = \frac{35}{5} - \frac{17}{5}$

$= \frac{35 - 17}{5} = \frac{18}{5} = 3\frac{3}{5}$

The estimate was 4, so an exact answer of $3\frac{3}{5}$ is reasonable.

c. $9 - 4\frac{1}{7}$

Estimate _____

Exact _____

Review these examples for Objective 5:

8. First, estimate the answer to each application problem. Then find the exact answer.

a. Thomas worked $25\frac{1}{2}$ hours over the last four days. If he worked the same amount each day, how long was he at work each day?

First, round each mixed number to the nearest whole number.

$25\frac{1}{2}$ rounds to 26

Using the rounded numbers, we divide.

$26 \div 4 = 6\frac{1}{2} \leftarrow$ Estimate

To find the exact answer, use the original mixed numbers and divide.

$25\frac{1}{2} \div 4 = \frac{51}{2} \div \frac{4}{1} = \frac{51}{2} \cdot \frac{1}{4} = \frac{51 \cdot 1}{2 \cdot 4} = \frac{51}{8} = 6\frac{3}{8}$

Thomas worked $6\frac{3}{8}$ hr each day. This is close to the estimate of $6\frac{1}{2}$ hr.

b. George's daughter grew $1\frac{1}{3}$ inches last year and $2\frac{1}{5}$ inches this year. How much has her height increased over the two years?

First, round each mixed number to the nearest whole number.

$1\frac{1}{3}$ rounds to 1   and $2\frac{1}{5}$ rounds to 2

Using the rounded numbers, we add.

$1 + 2 = 3 \leftarrow$ Estimate

Now Try:

8. First, estimate the answer to each application problem. Then find the exact answer.

a. Katlyn used $2\frac{3}{8}$ packages of nuts in her salad recipe. Each package has $3\frac{1}{6}$ ounces of nuts. How many ounces of nuts did she use in the recipe?

Estimate _____

Exact _____

b. A plumber has three pieces of pipe measuring $2\frac{1}{5}$ ft, $3\frac{3}{4}$ ft, and $4\frac{1}{8}$ ft. What is the total length of pipe?

Estimate _____

Exact _____

To find the exact answer, use the original mixed numbers and add.

$$1\frac{1}{3}+2\frac{1}{5}=\frac{4}{3}+\frac{11}{5}=\frac{20}{15}+\frac{33}{15}=\frac{20+33}{15}$$

$$=\frac{53}{15}=3\frac{8}{15}$$

Her height increased $3\frac{8}{15}$ in. over the two years.

This result is close to the estimate of 3 in.

Objective 1   Identify mixed numbers and graph them on a number line.

For extra help, see Example 1 on page 266 of your text and Section Lecture video for Section 4.5 and Exercise Solutions Clip 1.

Graph the mixed numbers or improper fractions on a number line.

1. $-3\frac{1}{2}$ and $3\frac{1}{2}$

     1.

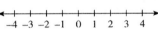

2. $\frac{10}{3}$ and $-\frac{10}{3}$

     2.

3. $-\frac{11}{4}$ and $\frac{11}{4}$

     3.

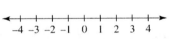

Objective 2   Rewrite mixed numbers as improper fractions, or the reverse.

For extra help, see Examples 2–3 on pages 267–269 of your text and Section Lecture video for Section 4.5 and Exercise Solutions Clip 23.

Write each mixed number as an improper fraction.

4. $-8\frac{2}{7}$

     4. _____

5. $-1\frac{7}{9}$

     5. _____

Write the improper fraction as a mixed number in simplest form.

6. $\frac{26}{3}$

     6. _____

Name: Date:
Instructor: Section:

Objective 3   Estimate the answer and multiply or divide mixed numbers.

For extra help, see Examples 4–6 on page 269–271 of your text and Section Lecture video for Section 4.5 and Exercise Solutions Clip 25 and 27.

First, round the mixed numbers to the nearest whole number and estimate each answer; then find the exact answer. Write exact answers in simplest form. Write each mixed number as an improper fraction.

7. $3\frac{2}{3} \cdot 1\frac{1}{7}$

7.
Estimate_____

Exact _____

8. $3\frac{3}{4} \div 1\frac{3}{8}$

8.
Estimate_____

Exact _____

9. $5\frac{1}{3} \div 2\frac{2}{5}$

9.
Estimate_____

Exact _____

Objective 4   Estimate the answer and add or subtract mixed numbers.

For extra help, see Example 7 on pages 272–273 of your text and Section Lecture video for Section 4.5 and Exercise Solutions Clip 29 and 31.

First, round the mixed numbers to the nearest whole number and estimate each answer; then find the exact answer. Write exact answers in simplest form. Write each mixed number as an improper fraction.

10. $3\frac{5}{6} + 2\frac{3}{4}$

10.
Estimate_____

Exact _____

11. $9 - 5\dfrac{4}{9}$

11.
Estimate_____

Exact _____

12. $9\dfrac{7}{18} - 3\dfrac{5}{12}$

12.
Estimate_____

Exact _____

Objective 5   Solve application problems containing mixed numbers.

For extra help, see Example 8 on page 274 of your text and Section Lecture video for Section 4.5 and Exercise Solutions Clip 49.

First, estimate the answer to each application problem. Then find the exact answer.

13. A living room has dimensions $3\dfrac{3}{4}$ meters by $3\dfrac{1}{3}$ meters. What is the area of the room?

13.
Estimate_____

Exact _____

14. Suppose that a pair of pants requires $3\dfrac{1}{8}$ yd of material. How much material would be needed for 6 pairs of pants?

14.
Estimate_____

Exact _____

Chapter 4 RATIONAL NUMBERS: POSITIVE AND NEGATIVE FRACTIONS

4.6   Exponents, Order of Operations, and Complex Fractions

Learning Objectives	
1	Simplify fractions with exponents.
2	Use the order of operations with fractions.
3	Simplify complex fractions.

Key Terms

Use the vocabulary terms listed below to complete each statement in exercises 1−3.

complex fraction   exponent   order of operations

1. For problems or expressions with more than one operation, the _____ tells what to do first, second, and so on, to obtain the correct answer.

2. An _____ tells how many times a number is used as a factor in repeated multiplication.

3. A _____ is a fraction in which the numerator and/or denominator contain one or more fractions.

Guided Examples

Review these examples for Objective 1:
1. Simplify.

a. $\left(-\dfrac{1}{5}\right)^3$

$\left(-\dfrac{1}{5}\right)^3 = \left(-\dfrac{1}{5}\right)\left(-\dfrac{1}{5}\right)\left(-\dfrac{1}{5}\right) = \dfrac{1}{25}\left(-\dfrac{1}{5}\right) = -\dfrac{1}{125}$

b. $\left(\dfrac{4}{5}\right)^3 \left(\dfrac{5}{8}\right)^2$

$\left(\dfrac{4}{5}\right)^3 \left(\dfrac{5}{8}\right)^2 = \left(\dfrac{4}{5}\right)\left(\dfrac{4}{5}\right)\left(\dfrac{4}{5}\right)\left(\dfrac{5}{8}\right)\left(\dfrac{5}{8}\right)$

$= \dfrac{\overset{1}{\cancel{2}}\cdot\overset{1}{\cancel{2}}\cdot\overset{1}{\cancel{4}}\cdot\overset{1}{\cancel{4}}\cdot\overset{1}{\cancel{8}}\cdot\overset{1}{\cancel{8}}}{\underset{1}{\cancel{8}}\cdot\underset{1}{\cancel{8}}\cdot 5\cdot\underset{1}{\cancel{2}}\cdot\underset{1}{\cancel{4}}\cdot\underset{1}{\cancel{2}}\cdot\underset{1}{\cancel{4}}}$

$= \dfrac{1}{5}$

Now Try:
1. Simplify.

a. $\left(-\dfrac{1}{4}\right)^3$

_____

b. $\left(\dfrac{5}{7}\right)^3 \left(\dfrac{7}{10}\right)^2$

_____

Review these examples for Objective 2:

2. Simplify.

    a. $-\dfrac{1}{2}+\dfrac{2}{3}\left(\dfrac{4}{5}\right)$

    $$-\dfrac{1}{2}+\dfrac{2}{3}\left(\dfrac{4}{5}\right)=-\dfrac{1}{2}+\dfrac{2\cdot 4}{3\cdot 5}=-\dfrac{1}{2}+\dfrac{8}{15}$$

    $$=-\dfrac{15}{30}+\dfrac{16}{30}=\dfrac{-15+16}{30}$$

    $$=\dfrac{1}{30}$$

    b. $-4+\left(\dfrac{5}{6}-\dfrac{2}{3}\right)^{2}$

    $$-4+\left(\dfrac{5}{6}-\dfrac{2}{3}\right)^{2}=-4+\left(\dfrac{5}{6}-\dfrac{4}{6}\right)^{2}$$

    $$=-4+\left(\dfrac{5-4}{6}\right)^{2}=-4+\left(\dfrac{1}{6}\right)^{2}$$

    $$=-4+\left(\dfrac{1}{36}\right)$$

    $$=\dfrac{-144+1}{36}=-\dfrac{143}{36}$$

Now Try:

2. Simplify.

    a. $-\dfrac{1}{6}+\dfrac{2}{3}\left(\dfrac{4}{5}\right)$

    _____

    b. $-1+\left(\dfrac{4}{9}-\dfrac{2}{3}\right)^{2}$

    _____

Review these examples for Objective 3:

3. Simplify.

    a. $\dfrac{-\dfrac{6}{7}}{-\dfrac{5}{14}}$

    Rewrite the complex fraction using the $\div$ symbol for division. Then follow the order of operations.

    $$\dfrac{-\dfrac{6}{7}}{-\dfrac{5}{14}}=-\dfrac{6}{7}\div-\dfrac{5}{14}=-\dfrac{6}{7}\cdot-\dfrac{14}{5}$$

    $$=\dfrac{6\cdot 2\cdot \overset{1}{\cancel{7}}}{\underset{1}{\cancel{7}}\cdot 5}=\dfrac{12}{5},\text{ or }2\dfrac{2}{5}$$

    b. $\dfrac{\left(\dfrac{5}{6}\right)^{2}}{10}$

    Rewrite the complex fraction using the $\div$

Now Try:

3. Simplify.

    a. $\dfrac{-\dfrac{5}{9}}{-\dfrac{7}{12}}$

    _____

    b. $\dfrac{\left(\dfrac{3}{5}\right)^{2}}{9}$

    _____

symbol for division. Then follow the order of operations.

$$\frac{\left(\frac{5}{6}\right)^2}{10} = \left(\frac{5}{6}\right)^2 \div 10 = \frac{25}{36} \div 10 = \frac{25}{36} \cdot \frac{1}{10}$$

$$= \frac{\overset{1}{\cancel{5}} \cdot 5 \cdot 1}{36 \cdot 2 \cdot \underset{1}{\cancel{5}}} = \frac{5}{72}$$

Objective 1   Simplify fractions with exponents.

For extra help, see Example 1 on page 281 of your text and Section Lecture video for Section 4.6 and Exercise Solutions Clip 7 and 13.

Simplify.

1. $3\left(-\frac{1}{3}\right)^3$

1. _____

2. $\left(-\frac{2}{3}\right)^3\left(-\frac{3}{2}\right)^2$

2. _____

3. $\left(-\frac{2}{3}\right)^2\left(\frac{1}{2}\right)^4$

3. _____

Objective 2   Use the order of operations with fractions.

For extra help, see Example 2 on page 282 of your text and Section Lecture video for Section 4.6 and Exercise Solutions Clip 23 and 33.

Simplify.

4. $\frac{15}{16} - \frac{1}{8} - \left(\frac{3}{4}\right)^2$

4. _____

5. $\left(\frac{3}{2}\right)^2 - \frac{3}{10} \div \frac{1}{5}$

5. _____

6. $\left(\frac{1}{3}\right)^2 - \left(-\frac{1}{2}\right)^3$

6. _____

Name:

Instructor:

Date:

Section:

Objective 3 Simplify complex fractions.

For extra help, see Example 3 on page 283 of your text and Section Lecture video for Section 4.6 and Exercise Solutions Clip 49.

Simplify.

7. $\dfrac{\dfrac{16}{35}}{-\dfrac{4}{70}}$

7. _____

8. $\dfrac{-20}{\dfrac{2}{5}}$

8. _____

9. _____

9. $\dfrac{\left(\dfrac{2}{3}\right)^2}{\left(-\dfrac{4}{5}\right)^2}$

Chapter 4 RATIONAL NUMBERS: POSITIVE AND NEGATIVE FRACTIONS

4.7   Problem Solving: Equations Containing Fractions

Learning Objectives

1   Use the multiplication property of equality to solve equations containing fractions.

2   Use both the addition and the multiplication properties of equality to solve equations containing fractions.

3   Solve application problems using equations containing fractions.

Key Terms

Use the vocabulary terms listed below to complete each statement in exercises 1−2.

multiplication property of equality

division property of equality

1.   The _____ states that both sides of an equation may be divided by the same nonzero number and it will still be balanced.

2.   The _____ states that both sides of an equation may be multiplied by the same nonzero number and it will still be balanced.

Guided Examples

Review these examples for Objective 1:

1.   Solve each equation and check each solution.

    a. $\dfrac{1}{5}b = 12$

Multiply both sides by $\dfrac{5}{1}$.

$$\frac{1}{5}b = 12$$

$$\frac{5}{1}\left(\frac{1}{5}b\right) = \frac{5}{1}(12)$$

$$\left(\frac{5}{1} \cdot \frac{1}{5}\right)b = \frac{5}{1}\left(\frac{12}{1}\right)$$

$$1b = \frac{60}{1}$$

$$b = 60$$

Now Try:

1.   Solve each equation and check each solution.

    a. $\dfrac{1}{7}b = 3$

_____

Check $\dfrac{1}{5}b = 12$

$\dfrac{1}{5}(60) = 12$

$\dfrac{1 \cdot \cancel{5} \cdot 12}{\cancel{5}} = 12$

$12 = 12$

The solution is 60.

b. $15 = -\dfrac{3}{5}x$

Multiply both sides by $-\dfrac{5}{3}$.

$-\dfrac{5}{3}(15) = -\dfrac{5}{3}\left(-\dfrac{3}{5}x\right)$

$-25 = x$

Check $15 = -\dfrac{3}{5}x$

$15 = -\dfrac{3}{5}(-25)$

$15 = \dfrac{3 \cdot \cancel{5} \cdot 5}{\cancel{5}}$

$15 = 15$

The solution is –25.

c. $-\dfrac{5}{6}n = -\dfrac{1}{4}$

$-\dfrac{6}{5}\left(-\dfrac{5}{6}n\right) = -\dfrac{6}{5}\left(-\dfrac{1}{4}\right)$

$n = \dfrac{2 \cdot 3}{5 \cdot 2 \cdot 2}$

$n = \dfrac{3}{10}$

Check $-\dfrac{5}{6}n = -\dfrac{1}{4}$

$-\dfrac{5}{6}\left(\dfrac{3}{10}\right) = -\dfrac{1}{4}$

$-\dfrac{\cancel{5} \cdot \cancel{3}}{2 \cdot \cancel{3} \cdot 2 \cdot \cancel{5}} = -\dfrac{1}{4}$

$-\dfrac{1}{4} = -\dfrac{1}{4}$

The solution is $\dfrac{3}{10}$.

b. $25 = -\dfrac{5}{6}x$

_____

c. $-\dfrac{7}{9}n = -\dfrac{14}{45}$

_____

Review these examples for Objective 2:

2. Solve each equation and check each solution.

a. $\frac{1}{5}c + 2 = 5$

Add the opposite of 2, which is –2, to both sides.

$$\frac{1}{5}c + 2 = 5$$
$$\underline{\quad -2 \quad -2}$$
$$\frac{1}{5}c + 0 = 3$$
$$\frac{1}{5}c = 3$$
$$\frac{5}{1}\left(\frac{1}{5}c\right) = \frac{5}{1}\left(\frac{3}{1}\right)$$
$$c = 15$$

Check $\frac{1}{5}c + 2 = 5$

$$\frac{1}{5}(15) + 2 = 5$$
$$3 + 2 = 5$$
$$5 = 5$$

The solution is 15.

b. $-5 = \frac{7}{8}y + 9$

To get the variable term by itself, add –9 to both sides.

$$-5 = \frac{7}{8}y + 9$$
$$\underline{-9 \qquad\quad -9}$$
$$-14 = \frac{7}{8}y + 0$$
$$\frac{8}{7}(-14) = \frac{8}{7}\left(\frac{7}{8}y\right)$$
$$-16 = y$$

Check $-5 = \frac{7}{8}y + 9$

$$-5 = \frac{7}{8}(-16) + 9$$
$$-5 = -14 + 9$$
$$-5 = -5$$

The solution is –16.

Now Try:

2. Solve each equation and check each solution.

a. $\frac{1}{6}c + 3 = 8$

_____

b. $-9 = \frac{5}{7}y + 6$

_____

Review this example for Objective 3:

3. An expression for the recommended weight of an adult is $\frac{11}{2}$ (height in inches) -220.

   A man weighs 187 pounds. What is his height in inches?

   Step 1 Read the problem. It is about weight and height.

   Step 2 Assign a variable. Let h represent the height.

   Step 3 Write an equation.

   $$\frac{11}{2}(\text{height in inches}) - 220 \text{ is weight}$$

   $$\frac{11}{2}h - 220 = 187$$

   Step 4 Solve.

   $$\frac{11}{2}h - 220 = 187$$

   $$\underline{+220 \quad +220}$$

   $$\frac{11}{2}h + 0 = 407$$

   $$\frac{2}{11}\left(\frac{11}{2}h\right) = \frac{2}{11}(407)$$

   $$h = 74$$

   Step 5 State the answer. The man is 74 inches, or 6 ft 2 in.

   Step 6 Check.
   If the height is 74 inches, then find the weight.

   $$\frac{11}{2}(74) - 220 = 407 - 220 = 187$$

   The answer checks.

Now Try:

3. An expression for the recommended weight of an adult is $\frac{11}{2}$ (height in inches) -220.

   A man weighs 220 pounds. What is his height in inches?

   _____

Objective 1   **Use the multiplication property of equality to solve equations containing fractions.**

For extra help, see Example 1 on pages 288–290 of your text and Section Lecture video for Section 4.7 and Exercise Solutions Clip 7.

Solve each equation and check each solution.

1. $-30 = \frac{5}{6}b$ 1. _____

2. $\frac{2}{9}n = 18$ 2. _____

3. $-\frac{8}{9}h = -\frac{1}{6}$ 3. _____

Objective 2 Use both the addition and the multiplication properties of equality to solve equations containing fractions.

For extra help, see Example 2 on pages 290–291 of your text and Section Lecture video for Section 4.7 and Exercise Solutions Clip 19 and 21.

Solve each equation and check each solution.

4. $\frac{1}{5}s - 15 = -10$ 4. _____

5. $-8 = \frac{5}{2}r + 2$ 5. _____

6. $-\frac{1}{6}n + 8 = -4$ 6. _____

Objective 3 Solve application problems using equations containing fractions.

For extra help, see Example 3 on page 292 of your text and Section Lecture video for Section 4.7.

In Exercises 7–9, find each person's age using the six problem-solving steps and this expression for approximate systolic blood pressure: $100 + \frac{age}{2}$. *Assume that all the people have normal blood pressure.*

7. A man has a systolic blood pressure of 110. How old is he? 7. _____

8. A woman has a systolic blood pressure of 126. How old is she? 8. _____

9. A man has a systolic blood pressure of 115. How old is he? 9. _____

Name: Date:
Instructor: Section:

Chapter 4 RATIONAL NUMBERS: POSITIVE AND NEGATIVE FRACTIONS

4.8   Geometry Applications: Area and Volume

Learning Objectives
1   Find the area of triangle.
2   Find the volume of a rectangular solid.
3   Find the volume of a pyramid.

Key Terms

Use the vocabulary terms listed below to complete each statement in exercises 1−2.

> **volume   area**

1. _____ is measured in square units.

2. _____ is measured in cubic units.

Guided Examples

Review these examples for Objective 1:

1. Find the area of each triangle.

    a.

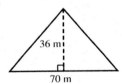

The base is 70 m and the height is 36 m.

$$A = \frac{1}{2} \cdot b \cdot h$$

$$A = \frac{1}{\underset{1}{2}} \cdot \overset{35}{\cancel{70}} \text{ m} \cdot 36 \text{ m}$$

$$A = 1260 \text{ m}^2$$

    b.

             15 ft

$13\frac{1}{8}$ ft

Two sides of the triangle are perpendicular to each other, so use those sides as the base and the height.

Now Try:

1. Find the area of each triangle.

    a.

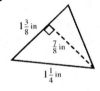

_____

    b.

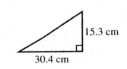

_____

$$A = \frac{1}{2} \cdot 15 \text{ ft} \cdot 13\frac{1}{8} \text{ ft}$$

$$A = \frac{1}{2} \cdot \frac{15 \text{ ft}}{1} \cdot \frac{105 \text{ ft}}{8}$$

$$A = 98\frac{7}{16} \text{ ft}^2$$

2. Find the area of the shaded part of this figure.

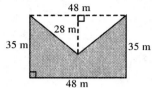

The entire figure is a rectangle. Find the area of the rectangle.

$$A = l \cdot w$$

$$A = 48 \text{ m} \cdot 35 \text{ m}$$

$$A = 1680 \text{ m}^2$$

The unshaded part is a triangle. Find the area of the triangle.

$$A = \frac{1}{2} \cdot b \cdot h$$

$$A = \frac{1}{\cancel{2}_{1}} \cdot \cancel{48}^{24} \text{ m} \cdot 28 \text{ m}$$

$$A = 672 \text{ m}^2$$

Subtract to find the area of the shaded part.

$$\overbrace{A = 1680 \text{ m}^2}^{\text{Entire area}} - \overbrace{672 \text{ m}^2}^{\text{Unshaded part}} = \overbrace{1008 \text{ m}^2}^{\text{Shaded part}}$$

The area of the shaded part is 1008 m².

Review these examples for Objective 2:

3. Find the volume of each box.

a.

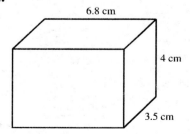

Use the formula $V = l \cdot w \cdot h$.

$$V = 6.8 \text{ cm} \cdot 3.5 \text{ cm} \cdot 4 \text{ cm}$$

$$V = 95.2 \text{ cm}^3$$

2. Find the area of the shaded part of this figure.

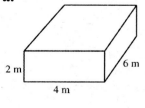

_____

Now Try:

3. Find the volume of each box.

a.

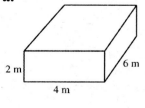

_____

b.

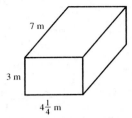

Use the formula $V = l \cdot w \cdot h$.

$$V = 7 \text{ m} \cdot 4\frac{1}{4} \text{ m} \cdot 3 \text{ m}$$

$$V = \frac{7 \text{ m}}{1} \cdot \frac{17 \text{ m}}{4} \cdot \frac{3 \text{ m}}{1}$$

$$V = \frac{7 \text{ m} \cdot 17 \text{ m} \cdot 3 \text{ m}}{4}$$

$$V = \frac{357}{4} \text{ m}^3, \text{ or } 89\frac{1}{4} \text{ m}^3$$

b.

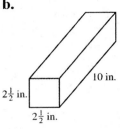

_____

Review this example for Objective 3:

4. Find the volume of this pyramid with rectangular base. Round your answer to the nearest tenth.

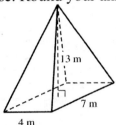

First find the value of B in the formula, which is the area of a rectangular base. Recall that the area of the rectangle is found by multiplying length times width.

$$B = 7 \text{ m} \cdot 4 \text{ m}$$

$$B = 28 \text{ m}^2$$

Now find the volume.

$$V = \frac{B \cdot h}{3}$$

$$V \approx \frac{28 \text{ m}^2 \cdot 13 \text{ m}}{3}$$

$$V \approx 121.3 \text{ m}^3$$

Now Try:

4. Find the volume of this pyramid with rectangular base. Round your answer to the nearest tenth.

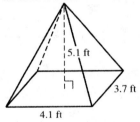

_____

Name: Date:

Instructor: Section:

Objective 1   Find the area of triangle.

For extra help, see Examples 1–2 on pages 298–299 of your text and Section Lecture video for Section 4.8 and Exercise Solutions Clip 5.

Find the perimeter and area of each triangle.

1.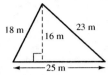

                                             1. Perimeter_____

                                              Area _____

2.

                                             2. Perimeter_____

                                              Area _____

Find the shaded area in the figure.

3.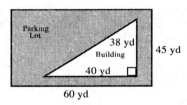

                                             3. _____

Objective 2   Find the volume of a rectangular solid.

For extra help, see Example 3 on page 300 of your text and Section Lecture video for Section 4.8 and Exercise Solutions Clip 17.

Find the volume of each figure.

4.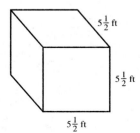

                                             4. _____

5.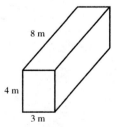

                                               5. _____

Name: Date:
Instructor: Section:

6.
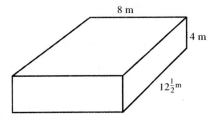

6. _____

Objective 3   Find the volume of a pyramid.

For extra help, see Example 4 on page 301 of your text and Section Lecture video for Section 4.8 and Exercise Solutions Clip 21 and 27.

Find the volume of each figure.

7.
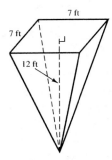

7. _____

8.
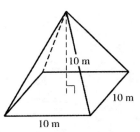

8. _____

9.
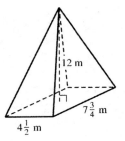

9. _____

Chapter 5 RATIONAL NUMBERS: POSITIVE AND NEGATIVE DECIMALS

5.1   Reading and Writing Decimals

Learning Objectives
1   Write parts of a whole using decimals.
2   Identify the place value of a digit.
3   Read decimal numbers.
4   Write decimals as fractions or mixed numbers.

Key Terms

Use the vocabulary terms listed below to complete each statement in exercises 1–3.

decimals   **decimal point**   **place value**

1. We use _____ to show parts of a whole.

2. A _____ is assigned to each place to the left or right of the decimal point.

3. The dot that separates the whole number part from the fractional part of a decimal number is called the _____.

Guided Examples

Review these examples for Objective 1:

1. Given a fraction and how it is read, write as a decimal.

    a. $\frac{7}{10}$ seven tenths

    Decimal: 0.7

    b. $\frac{3}{100}$ three hundredths

    Decimal: 0.03

    c. $\frac{63}{100}$ sixty-three hundredths

    Decimal: 0.63

Now Try:

1. Given a fraction and how it is read, write as a decimal.

    a. $\frac{6}{10}$ six tenths

    _____

    b. $\frac{5}{100}$ five hundredths

    _____

    c. $\frac{37}{100}$ thirty-seven hundredths

    _____

d. $\dfrac{9}{1000}$ nine thousandths

Decimal: 0.009

e. $\dfrac{56}{1000}$ fifty-six thousandths

Decimal: 0.056

f. $\dfrac{943}{1000}$ nine hundred forty-three thousandths

Decimal: 0.943

d. $\dfrac{7}{1000}$ seven thousandths

_____

e. $\dfrac{49}{1000}$ forty-nine thousandths

_____

f. $\dfrac{518}{1000}$ five hundred eighteen thousandths

_____

Review these examples for Objective 2:

2. Identify the place value of each digit.

    a. 486.92

    4 hundreds
    8 tens
    6 ones
    .
    9 tenths
    2 hundredths

    b. 0.00465

    0 ones
    .
    0 tenths
    0 hundredths
    4 thousandths
    6 ten-thousandths
    5 hundred-thousandths

Now Try:

2. Identify the place value of each digit.

    a. 862.93

    _____

    b. 0.00769

    _____

Review these examples for Objective 3:

3. Tell how to read each decimal in words.

    a. 0.7

    Read it as: seven tenths

    b. 0.62

    Read it as: sixty-two hundredths

    c. 0.09

    Read it as: nine hundredths

Now Try:

3. Tell how to read each decimal in words.

    a. 0.9

    _____

    b. 0.53

    _____

    c. 0.07

    _____

d. 0.352

Read it as: three hundred fifty-two thousandths

e. 0.0205

Read it as: two hundred five ten-thousandths

4. Read each decimal.

   a. 17.8

seventeen and eight tenths

   b. 543.71

five hundred forty-three and seventy-one hundredths

   c. 0.089

eighty-nine thousandths

   d. 12.4053

twelve and four thousand fifty-three ten-thousandths

d. 0.502

_____

e. 0.0469

_____

4. Read each decimal.

   a. 4.7

_____

   b. 18.009

_____

   c. 0.0082

_____

   d. 57.906

_____

Review these examples for Objective 4:

5. Write each decimal as a fraction or mixed number.

   a. 0.16

The digits to the right of the decimal point, 16, are the numerator of the fraction. The denominator is 100 for hundredths because the rightmost digit is in the hundredths place.

$$0.16 = \frac{16}{100}$$

   b. 0.739

The rightmost digit is in the thousandths place.

$$0.739 = \frac{739}{1000}$$

   c. 5.0088

The whole number part stays the same. The rightmost digit is in the ten-thousandths place.

$$5.0088 = 5\frac{88}{10,000}$$

Now Try:

5. Write each decimal as a fraction or mixed number.

   a. 0.27

_____

   b. 0.303

_____

   c. 3.2636

_____

Name: Date:

Instructor: Section:

6. Write each decimal as a fraction or mixed number in lowest terms.

a. 0.6

$$0.6 = \frac{6}{10} \qquad \text{Write } \frac{6}{10} \text{ in lowest terms.}$$

$$\frac{6}{10} = \frac{6 \div 2}{10 \div 2} = \frac{3}{5}$$

b. 0.88

$$0.88 = \frac{88}{100} = \frac{88 \div 4}{100 \div 4} = \frac{22}{25}$$

c. 17.216

$$17.216 = 17\frac{216}{1000} = 17\frac{216 \div 8}{1000 \div 8} = 17\frac{27}{125}$$

d. 53.6065

$$53.6065 = 53\frac{6065}{10,000} = 53\frac{6065 \div 5}{10,000 \div 5} = 53\frac{1213}{2000}$$

6. Write each decimal as a fraction or mixed number in lowest terms.

a. 0.2

_____

b. 0.35

_____

c. 6.04

_____

d. 562.0404

_____

Objective 1 Write parts of a whole using decimals.

For extra help, see Example 1 on page 323 of your text and Section Lecture video for Section 5.1.

Write the portion of each square that is shaded as a fraction, as a decimal, and in words.

1.

1. _____

_____

_____

2.

2. _____

_____

_____

3.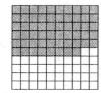

3. _____

_____

_____

Name: Date:

Instructor: Section:

Objective 2   Identify the place value of a digit.

For extra help, see Example 2 on page 324 of your text and Section Lecture video for Section 5.1 and Exercise Solutions Clip 7 and 15.

Identify the digit that has the given place value.

4.   43.507   tenths   **4.** _____

                             hundredths   _____

5.   2.83714   thousandths   **5.** _____

                             ten-thousandths   _____

Identify the place value of each digit in these decimals.

6.   37.082   3   **6.** _____

                             7   _____

                             0   _____

                             8   _____

                             2   _____

Objective 3   Read decimal numbers.

For extra help, see Examples 3–4 on pages 324–325 of your text and Section Lecture video for Section 5.1 and Exercise Solutions Clip 41 and 43.

Tell how to read each decimal in words.

7.   0.08   **7.** _____

8.   10.835   **8.** _____

9.   97.008   **9.** _____

Objective 4   Write decimals as fractions or mixed numbers.

For extra help, see Examples 5–6 on page 326 of your text and Section Lecture video for Section 5.1 and Exercise Solutions Clip 21 and 29.

Write each decimal as a fraction or mixed number in lowest terms.

10.   0.001   **10.** _____

11.   3.6   **11.** _____

12.   0.95   **12.** _____

Chapter 5 RATIONAL NUMBERS: POSITIVE AND NEGATIVE DECIMALS

5.2   Rounding Decimal Numbers

Learning Objectives
1   Learn the rules for rounding decimals.
2   Round decimals to any given place.
3   Round money amounts to the nearest cent or nearest dollar.

Key Terms

Use the vocabulary terms listed below to complete each statement in exercises 1–2.

rounding   **decimal places**

1.   _____ are the number of digits to the right of the decimal point.

2.   When we "cut off" a number after a certain place value, we are _____
that number.

Guided Examples

Review these examples for Objective 2:

1.   Round 16.87453 to the nearest thousandth.

Step 1 Draw a "cut-off" line after the thousandths place.
16.874 | 53

Step 2 Look only at the first digit you are cutting off. Ignore the other digits you are cutting off.
16.874 | 53   (Ignore the 3)

Step 3 If the first digit you are cutting off is 5 or more, round up the part of the number you are keeping.

$$\begin{array}{r} 16.874\ |\ 53 \\ +\ 0.001 \\ \hline 16.875 \end{array}$$

So, 16.87453 rounded to the nearest thousandth is 16.875. We can write $16.87453 \approx 16.875$.

2.   Round to the place indicated.

a. 6.4387 to the nearest tenth

Step 1 Draw a cut-off line after the tenths place.
6.4 | 387

Step 2 Look only at the 3.
6.4 | 387   (Ignore the 8 and 7)

Now Try:

1.   Round 43.80290 to the nearest thousandth.

_____

2.   Round to the place indicated.

a. 0.7976 to the nearest hundredth

_____

Step 3 The first digit is 4 or less, so the part you are keeping stays the same.

6.4 | 387
—————
6.4

Rounding 6.4387 to the nearest tenth is 6.4. We can write $6.4387 \approx 6.4$.

b. 0.79846 to the nearest hundredth

Step 1 Draw a cut-off line after the hundredths place.

0.79 | 846

Step 2 Look only at the 8.

0.79 | 846

Step 3 The first digit is 5 or more, so round up by adding 1 hundredth to the part you are keeping.

$$\begin{array}{r} 1 \\ 0.79 \mid 846 \\ +\ 0.01 \\ \hline 0.80 \end{array}$$

0.79846 rounded to the nearest hundredth is 0.80. We can write $0.79846 \approx 0.80$.

c. 0.02709 to the nearest thousandth

0.027 | 09

The first digit cut is 4 or less, so the part you are keeping stays the same.

0.02709 rounded to the nearest thousandth is 0.027. We can write $0.02709 \approx 0.027$.

d. 64.983 to the nearest tenth

64.9 | 83

The first digit cut is 5 or more, so round up by adding 1 tenth to the part you are keeping.

$$\begin{array}{r} 1 \\ 64.9 \mid 83 \\ +\ 0.1 \\ \hline 65.0 \end{array}$$

64.983 rounded to the nearest tenth is 65.0. We can write $64.983 \approx 65.0$. You must write the 0 in the tenths place to show that the number was rounded to the nearest tenth.

b. 7.7804 to the nearest hundredth

———————————————

c. 22.0397 to the nearest thousandth

———————————————

d. 0.649 to the nearest tenth

———————————————

Review these examples for Objective 3:

3. Round each money amount to the nearest cent.

a. $6.5348

Is $6.5348 closer to $6.53 or to $6.54?
The first digit cut is 4 or less, so the part you are keeping stays the same.
    $6.53 | 48
You pay $6.53.

b. $0.895

Is $0.895 closer to $0.89 or $0.90?
    $0.89 | 5
The first digit cut is 5 or more, so round up.

$$\begin{array}{r} 1 \\ \$0.89 \mid 5 \\ +\ \$0.01 \\ \hline \$0.90 \end{array}$$

You pay $0.90.

4. Round to the nearest dollar.

a. $56.76

Draw a cut-off line after the ones place.
    $56|.76
First digit cut is 5 or more, so round up by adding $1.

$$\begin{array}{r} \$56 \mid .76 \\ +\qquad 1 \\ \hline \$57 \end{array}$$

So, $56.76 rounded to the nearest dollar is $57.

b. $697.41

Draw a cut-off line after the ones place.
    $697|.41
First digit cut is 4 or less, so the part you keep stays the same.
So, $697.41 rounded to the nearest dollar is $697.

c. $599.66

Draw a cut-off line after the ones place.
    $599|.66
First digit cut is 5 or more, so round up by adding $1.

Now Try:

3. Round each money amount to the nearest cent.

a. $2.0849

_____

b. $425.0954

_____

4. Round to the nearest dollar.

a. $37.81

_____

b. $307.20

_____

c. $880.83

_____

$599\,|\,.66$
$+\qquad 1$
$\overline{\$600}$

So, $599.66 rounded to the nearest dollar is
$600.

d. $5779.50

    $5779\,|\,.50$

First digit cut is 5 or more, so round up by
adding $1.

    $5779\,|\,.50$
$+\qquad 1$
$\overline{\$5780}$

$5779.50 rounded to the nearest dollar is $5780.

e. $0.53

    $0\,|\,.53$

First digit cut is 5 or more, so round up.
$0.53 rounded to the nearest dollar is $1.

d. $6859.77

    _____

e. $0.61

    _____

Objective 1   Learn the rules for rounding decimals.

For extra help, see page 331 of your text and Section Lecture video for Section 5.2.

Select the phrase that makes the sentence correct.

1.   When rounding a number to the nearest tenth, if the
digit in the hundredths place is 5 or more, round the
digit in the tenths place (up/down).

1. _____

2.   When rounding a number to the nearest hundredth,
look at the digit in the (tenth/thousandth) place.

2. _____

Objective 2   Round decimals to any given place.

For extra help, see Examples 1–2 on pages 331–333 of your text and Section Lecture
video for Section 5.2 and Exercise Solutions Clip 7, 13, and 17.

Round each number to the place indicated.

3.   489.84 to the nearest tenth

3. _____

4.   54.4029 to the nearest hundredth

4. _____

5.   989.98982 to the nearest thousandth

5. _____

Objective 3   Round money amounts to the nearest cent or nearest dollar.

For extra help, see Examples 3–4 on pages 334–335 of your text and Section Lecture video for Section 5.2 and Exercise Solutions Clip 21, 23, 31, and 33.

Round to the nearest dollar.

   6. $28.39   **6.** _____

   7. $11,839.73   **7.** _____

Round to the nearest cent.

   8. $1028.6666   **8.** _____

Chapter 5 RATIONAL NUMBERS: POSITIVE AND NEGATIVE DECIMALS

5.3   Adding and Subtracting Signed Decimal Numbers

Learning Objectives
1   Add and subtract positive decimals.
2   Add and subtract negative decimals.
3   Estimate the answer when adding or subtracting decimals.

Key Terms

Use the vocabulary terms listed below to complete each statement in exercises 1–2.

estimating   **front end rounding**

1.   With _____, we round to the highest possible place.

2.   Avoid common errors in working decimal problems by _____ the answer first.

Guided Examples

Review these examples for Objective 1:

1.   Find each sum.

    a. 29.73 and 56.84

Step 1 Write the numbers in columns with the decimal points lined up.

$$
\begin{array}{r}
2\,9.7\,3 \\
+\,5\,6.8\,4 \\
\end{array}
$$

Step 2 Add as if these were whole numbers.
Step 3 Line up decimal point in answer under the decimal points in problem.

$$
\begin{array}{r}
{\scriptstyle 1\ 1} \\
2\,9.7\,3 \\
+\,5\,6.8\,4 \\
\hline
8\,6.5\,7 \\
\end{array}
$$

    b. $8.437 + 5.361 + 13.295$

Write the numbers vertically with decimal points lined up. Then add.

$$
\begin{array}{r}
{\scriptstyle 1\ 1\ \ 1\ 1} \\
8.4\,3\,7 \\
5.3\,6\,1 \\
+\,1\,3.2\,9\,5 \\
\hline
2\,7.0\,9\,3 \\
\end{array}
$$

Now Try:

1.   Find each sum.

    a. $12.687 + 2.943$

    _____

    b. $0.428 + 16.005 + 5.276$

    _____

2. Find each sum.

   a. 6.7 + 0.41

There are two decimal places in 0.41, so write a 0 in the hundredths place in 6.7 so it has two decimal places also.

```
   6.70
 +0.41
 ------
   7.11
```

   b. 12 + 9.36 + 3.754

Write in zeros so that all the addends have three decimal places.

```
 12.000
  9.360
+ 3.754
-------
 25.114
```

3. Find each difference.

   a. 14.32 from 36.74

Step 1 Line up decimal points.

```
  36.74
 -14.32
```

Step 2 Both numbers have two decimal places; no need to write in zeros.

Step 3 Line up decimal point in answer.

```
  36.74
 -14.32
 ------
  22.42
```

   b. 167.53 minus 69.85

Regrouping is needed here.

```
 0 15 16 14 13
 1 6 7 . 5 3
 -   6 9 . 8 5
 ------------
     9 7 . 6 8
```

Line up decimal point in answer.

4. Find each difference.

   a. 19.6 from 38.264

Line up decimal points and write in zeros so both numbers have three decimal places.

```
  38.264
 -19.600
 -------
  18.664
```

2. Find each sum.

   a. 7.53 + 29.314

         _____

   b. 0.631 + 999.3 + 14

         _____

3. Find each difference.

   a. 7.352 from 18.964

         _____

   b. 50.43 minus 39.86

         _____

4. Find each difference.

   a. 3.87 from 8.524

         _____

b. $28.8 - 19.963$

Write in two zeros and subtract as usual.

$$\begin{array}{r} 28.800 \\ -\ 19.963 \\ \hline 8.837 \end{array}$$

c. 16 less 7.54

Write a decimal point and two zeros after 16. Subtract as usual.

$$\begin{array}{r} 16.00 \\ -\ \ 7.54 \\ \hline 8.46 \end{array}$$

b. $20 - 16.74$

_____

c. 1 less 0.499

_____

Review these examples for Objective 2:

5. Find each sum.

a. $-5.8 + (-21)$

Both addends are negative, so the sum will be negative. To begin, $|-5.8|$ is 5.8, and $|-21|$ is 21. Then add the absolute values.

$$\begin{array}{r} 5.8 \\ +\ 21.0 \\ \hline 26.8 \end{array} \leftarrow \text{Write in a decimal point and one 0}$$

$-5.8 + (-21) = -26.8$

b. $-4.36 + 0.973$

The addends have different signs. To begin, $|-4.36|$ is 4.36, and $|0.973|$ is 0.973. Then subtract the lesser absolute value from the greater.

$$\begin{array}{r} 4.360 \\ -\ 0.973 \\ \hline 3.387 \end{array} \leftarrow \text{Write one 0}$$

$-4.36 + 0.973 = -3.387$

6. Find each difference.

a. $5.4 - 13.89$

Rewrite subtractions as adding the opposite.

$5.4 - 13.89$

$5.4 + (-13.89)$

-13.89 has the greater absolute value and is negative, so the answer will be negative.

$5.4 + (-13.89) = -8.49$ $\begin{array}{r} 13.89 \\ -\ \ 5.40 \\ \hline 8.49 \end{array}$

Now Try:

5. Find each sum.

a. $-6.4 + (-32)$

_____

b. $-13.57 + 2.984$

_____

6. Find each difference.

a. $7.6 - 19.34$

_____

b. $-2.74-(-5.9)$

Rewrite subtractions as adding the opposite.
$-2.74-(-5.9)$

$-2.74+5.9$
5.9 has the greater absolute value and is positive, so the answer will be positive.
$-2.74-(-5.9)=3.16$

$$\begin{array}{r} 5.90 \\ -2.74 \\ \hline 3.16 \end{array}$$

c. $17.4-(1.58+0.63)$

Work the parentheses first.
$17.4-(1.58+0.63)$

$17.4-(2.21)$

$17.4+(-2.21)$

15.19

b. $-5.29-(-8.6)$

_____

c. $21.8-(4.37+0.89)$

_____

Review these examples for Objective 3:

7. Use front end rounding to round each number. Then add or subtract the rounded numbers to get an estimate answer. Finally, find the exact answer.

    a. Find the sum of 295.8 and 7.894.

    Estimate: _Exact:_

    $$\begin{array}{r} 300 \\ +8 \\ \hline 308 \end{array} \xleftarrow{\text{Rounds to}} \begin{array}{r} 295.800 \\ +7.894 \\ \hline 303.694 \end{array}$$

    The estimate goes out to the hundreds place and so does the exact answer. Therefore, the decimal point is probably in the correct place in the exact answer.

    b. Subtract $16.95 from $78.23.

    Estimate: _Exact:_

    $$\begin{array}{r} \$80 \\ -20 \\ \hline \$60 \end{array} \xleftarrow{\text{Rounds to}} \begin{array}{r} \$78.23 \\ -16.95 \\ \hline \$61.28 \end{array}$$

Now Try:

7. Use front end rounding to round each number. Then add or subtract the rounded numbers to get an estimate answer. Finally, find the exact answer.
    a. Find the sum of 5.74 and 8.107.

    _____

    _____

    b. Subtract $23.49 from $65.04.

    _____

    _____

c. $-2.753 - 8.4$

Rewrite the subtraction as adding the opposite. Then use front end rounding to get an estimated answer.

$$-2.753 - 8.4$$
$$-2.753 + (-8.4)$$

Rounded $\rightarrow$ -3 $+$ $(-8) = -11$

To find the exact answer, add the absolute values.

$$\begin{array}{r} 2.753 \\ +\ 8.400 \\ \hline 11.153 \end{array}$$

The answer will be negative because both numbers are negative.

$$-2.753 - 8.4 = -11.153$$

The exact answer -11.153 is reasonable because it is close to the estimated answer of -11.

c. $-1.497 - 9.6$

_____

_____

Objective 1   Add and subtract positive decimals.

For extra help, see Examples 1–4 on pages 338–340 of your text and Section Lecture video for Section 5.3 and Exercise Solutions Clip 7, 11, and 23.

Find each sum or difference.

1. $43.96 + 48.53$

1. _____

2. $69.524 - 26.958$

2. _____

3. $45.83 + 20.923 + 5.7$

3. _____

Objective 2   Add and subtract negative decimals.

For extra help, see Examples 5–6 on pages 341–342 of your text and Section Lecture video for Section 5.3 and Exercise Solutions Clip 31 and 33.

Find each difference.

4. $18.1 - 84.6$

4. _____

5. $87.6 - (-90.4)$ **5.** _____

6. $1.71 - 12.68$ **6.** _____

Objective 3   Estimate the answer when adding or subtracting decimals.

For extra help, see Example 7 on pages 342–343 of your text and Section Lecture video for Section 5.3.

First, use front end rounding and estimate each answer. Then add or subtract to find the exact answer.

7.   593.8   **7.**
          27.93   **Estimate**_____
        $+ 54.87$ **Exact** _____

8. $7.69 - 20.85$ **8.**
                                      Estimate_____
                                      Exact _____

9. $-9.7 - 4.862$ **9.**
                                      Estimate_____
                                      Exact _____

Chapter 5 RATIONAL NUMBERS: POSITIVE AND NEGATIVE DECIMALS

5.4   Multiplying Signed Decimal Numbers

Learning Objectives
1   Multiply positive and negative decimals.
2   Estimate the answer when multiplying decimals.

Key Terms

Use the vocabulary terms listed below to complete each statement in exercises 1−3.

        decimal places   **factor**   **product**

1.   Each number in a multiplication problem is called a _____.

2.   When multiplying decimal numbers, first multiply the numbers, then find the total number of _____ in both factors.

3.   The answer to a multiplication problem is called the _____.

Guided Examples

Review this example for Objective 1:

1.   Find the product of 6.23 and −5.4.

    Step 1 Multiply the numbers as if they were whole numbers.

$$
\begin{array}{r}
6.23 \\
\times \quad 5.4 \\
\hline
2\,4\,9\,2 \\
3\,1\,1\,5 \\
\hline
3\,3\,6\,4\,2 \\
\end{array}
$$

    Step 2 Count the total number of decimal places in both factors.

$$
\begin{array}{r}
6.23 \leftarrow 2 \text{ decimal places} \\
\times \quad 5.4 \leftarrow 1 \text{ decimal place} \\
\hline
2\,4\,9\,2 \quad 3 \text{ total decimal places} \\
3\,1\,1\,5 \\
\hline
3\,3\,6\,4\,2 \\
\end{array}
$$

    Step 3 Count over 3 places in the product and write the decimal point. Count from right to left.

Now Try:

1.   Find the product of 2.51 and −4.3.

    _____

$$6.\,2\,3 \leftarrow 2 \text{ decimal places}$$
$$\underline{\times \qquad 5.\,4} \leftarrow 1 \text{ decimal place}$$
$$\overline{2\,4\,9\,2} \qquad 3 \text{ total decimal places}$$
$$\underline{3\,1\,1\,5}$$
$$\overline{3\,3.\,6\,4\,2}$$

Step 4 The factors have different signs, so the product is negative. 6.23 times –5.4 is –33.642.

2. Find the product: $(-0.035)(-0.07)$.

   Start by multiply, then count decimal places.
   $$0.\,0\,3\,5 \leftarrow 3 \text{ decimal places}$$
   $$\underline{\times \quad 0.\,0\,7} \leftarrow 2 \text{ decimal places}$$
   $$\overline{2\,4\,5} \qquad 5 \text{ total decimal places}$$

   After multiplying, the answer has only three decimal places, but five are needed, so write two zeros on the left side of the answer. Then count over 5 places and write in the decimal point.
   $$0.\,0\,3\,5 \leftarrow 3 \text{ decimal places}$$
   $$\underline{\times \quad 0.\,0\,7} \leftarrow 2 \text{ decimal places}$$
   $$\overline{.0\,0\,2\,4\,5} \qquad 5 \text{ total decimal places}$$

   The final product is 0.00245, which has five decimal places. The product is positive because the factors have the same sign.

2. Find the product $(-0.062)(-0.03)$.

   _____

Review this example for Objective 2:

3. First estimate the answer to $(74.56)(18.9)$ using front end rounding. Then find the exact answer.

   Estimate:
   $$70 \underleftarrow{\text{Rounds to}}$$
   $$\underline{\times 20 \underleftarrow{\text{Rounds to}}}$$
   $$\overline{1400}$$

   Exact:
   $$7\,4.\,5\,6 \leftarrow 2 \text{ decimal places}$$
   $$\underline{\times \qquad 1\,8.9} \leftarrow 1 \text{ decimal place}$$
   $$\overline{6\,7\,1\,0\,4} \qquad 3 \text{ total decimal places}$$
   $$5\,9\,6\,4\,8$$
   $$\underline{7\,4\,5\,6}$$
   $$\overline{1\,4\,0\,9.\,1\,8\,4}$$

   Both the estimate and the exact answer go out to the thousands place, so the decimal point in 1409.184 is probably in the correct place.

Now Try:

3. First estimate the answer to $(22.43)(5.03)$ using front end rounding. Then find the exact answer.

   _____

Name: _____   Date: _____

Instructor: _____   Section: _____

Objective 1   Multiply positive and negative decimals.

For extra help, see Examples 1–2 on pages 348–349 of your text and Section Lecture video for Section 5.4 and Exercise Solutions Clip 5, 7, and 19.

Find each product.

1. -19.3
      $\times\ 4.7$

1. _____

2. 0.682
      $\times\ \ 3.9$

2. _____

3. $(-0.074)(-0.05)$

3. _____

Objective 2   Estimate the answer when multiplying decimals.

For extra help, see Example 3 on page 350 of your text and Section Lecture video for Section 5.4 and Exercise Solutions Clip 23.

First use front-end rounding and estimate the answer. Then multiply to find the exact answer.

4. 29.8
      $\times\ 3.4$

4.
Estimate _____

Exact _____

5. 32.53
      $\times\ 23.26$

5.
Estimate _____

Exact _____

6. 391.9
      $\times\ \ 7.74$

6.
Estimate _____

Exact _____

Chapter 5 RATIONAL NUMBERS: POSITIVE AND NEGATIVE DECIMALS

5.5   Dividing Signed Decimal Numbers

Learning Objectives
1	Divide a decimal by an integer.
2	Divide a number by a decimal.
3	Estimate the answer when dividing decimals.
4	Use the order of operations with decimals.

Key Terms

Use the vocabulary terms listed below to complete each statement in exercises 1−4.

repeating decimal   quotient   dividend   divisor

1. In a division problem, the number being divided is called the _____.

2. The number $0.8\overline{3}$ is an example of a _____.

3. The answer to a division problem is called the _____.

4. In the problem $6.39 \div 0.9$, 0.9 is called the _____.

Guided Examples

Review these examples for Objective 1:
1. Find each quotient. Check the quotients by multiplying.

    a. $24.48 \div (-4)$

    First consider $24.48 \div 4$.
    Rewrite the division problem.
    Step 1 Write the decimal point in the quotient directly above the decimal point in the dividend.

    $$4\overline{)24.48}$$

    Step 2 Divide as if the numbers were whole numbers.

    $$\overset{6.12}{4\overline{)24.48}}$$

    Check by multiplying the quotient times the divisor.

Now Try:
1. Find each quotient. Check the quotients by multiplying.

    a. $9.891 \div (-7)$

    _____

$$6.12$$
$$\times \quad 4$$
$$\overline{24.48}$$

Step 3 The quotient is –6.12 because the numbers have different signs.

$$24.48 \div (-4) = -6.12$$

b. $8\overline{)425.6}$

Write the decimal point in the quotient above the decimal point in the dividend. Then divide as if the number were whole numbers.

$$
\begin{array}{r}
53.2 \\
8\overline{)425.6} \\
\underline{40} \\
25 \\
\underline{24} \\
16 \\
\underline{16} \\
0
\end{array}
$$
Check:
$$
\begin{array}{r}
53.2 \\
\times \quad 8 \\
\hline
425.6
\end{array}
$$

The quotient is 53.2.

2. Divide 1.35 by 4. Check the quotient by multiplying.

Divide.
$$
\begin{array}{r}
0.33 \\
4\overline{)1.35} \\
\underline{1\,2} \\
15 \\
\underline{12} \\
3
\end{array}
$$

Write a 0 after the 5 in the dividend so you can continue dividing. Keep writing more zeros in the dividend, if needed.

$$
\begin{array}{r}
0.3375 \\
4\overline{)1.3500} \\
\underline{1\,2} \\
15 \\
\underline{12} \\
30 \\
\underline{28} \\
20 \\
\underline{20} \\
0
\end{array}
$$
Check:
$$
\begin{array}{r}
0.3375 \\
\times \quad 4 \\
\hline
1.3500
\end{array}
$$

The quotient is 0.3375.

b. $5\overline{)75.15}$

_____

2. Divide 1008.9 by 50. Check the quotient by multiplying.

_____

3. Divide 8.87 by 9. Round the quotient to the nearest thousandth.

Write extra zeros in the dividend so you can continue dividing.

$$
\begin{array}{r}
0.9855 \\
9\overline{)8.8700} \\
\underline{8\ 1} \\
77 \\
\underline{72} \\
50 \\
\underline{45} \\
50 \\
\underline{45} \\
5
\end{array}
$$

Notice that the digit 5 in the answer is repeating. There are two ways to show that the answer is a repeating decimal that goes on forever.

    0.9855...   or $0.98\overline{5}$

To round to thousandths, divide out one more place, to ten-thousandths.

    $8.87 \div 9 = 0.9855...$ rounds to   0.986.

Check the answer by multiplying 0.986 by 9. Because 0.986 is a rounded answer, the check will not give exactly 8.87, but it should be very close.

    $(0.986)(9) = 8.874$

3. Divide 302.24 by 18. Round the quotient to the nearest thousandth.

_____

Review these examples for Objective 2:

4.

a. $\dfrac{41.2}{0.005}$

Move the decimal point in the divisor three places to the right so 0.005 becomes the whole number 5. Move the decimal point in the dividend the same number of places and write in two extra 0s.

$$
\begin{array}{r}
8240. \\
5\overline{)41200.}
\end{array}
$$

Now Try:

4.

a. $0.0024\overline{)48.984}$

_____

b. Divide –7 by –2.4. Round to the nearest hundredth.

First consider $7 \div 2.4$. Move the decimal point in the divisor one place to the right so 2.4 becomes the whole number 24. The decimal point in the dividend starts on the right side of 7 and is also moved one place to the right. (Remember, in order to round to hundredths, divide out one more place, to thousandths.)

```
        2.916
    24)70.000
       48
       220
       216
        40
        24
       160
       144
        16
```

Round the quotient. The quotient is positive because both the divisor and the dividend have the same sign.

$$-7 \div (-2.4) \approx 2.92$$

b. Divide –8 by –4.5. Round to the nearest hundredth.

_____

Review this example for Objective 3:

5. First use front end rounding to round each number and estimate the answer. Then divide to find the exact answer.

$$767.38 \div 3.7$$

Estimate:
```
     200
   4)800
```

Exact:
```
         27.4
    37)7673.8
       74
       273
       259
       148
       148
         0
```

Notice that the estimate, which is in hundreds, is very different from the exact answer, which is in tens.

   Find the error and rework.

The exact answer is 207.4, which fits the estimate of 200.

Now Try:

5. First use front end rounding to round each number and estimate the answer. Then divide to find the exact answer.

$$185.22 \div 4.9$$

_____

Review these examples for Objective 4:

6. Use the order of operations to simplify each expression.

    a. $4.5 + (-8.2)^2 + 11.63$

    Apply the exponent.
    $\quad 4.5 + 67.24 + 11.63$
    Add from left to right.
    $\quad 71.74 + 11.63$
    $\quad 83.37$

    b. $1.73 + (2.9 - 3.8)(6.5)$

    Work inside the parentheses.
    $\quad 1.73 + (-0.9)(6.5)$
    Multiply.
    $\quad 1.73 + (-5.85)$
    Add.
    $\quad -4.12$

    c. $5.6^2 - 1.4 \div 7(2.4)$

    Apply the exponent.
    $\quad 31.36 - 1.4 \div 7(2.4)$
    Multiply and divide from left to right.
    $\quad 31.36 - 0.2(2.4)$
    Multiply.
    $\quad 31.36 - 0.48$
    Subtract last.
    $\quad 30.88$

Now Try:

6. Use the order of operations to simplify each expression.

    a. $5.9 - 0.48 + (-2.6)^2$

    _____

    b. $4.06 \div 1.4 \times 7.8$

    _____

    c. $0.07 + 0.3(6.99 - 8)$

    _____

Objective 1   Divide a decimal by an integer.

For extra help, see Examples 1–3 on pages 355–357 of your text and Section Lecture video for Section 5.5 and Exercise Solutions Clip 5.

Find each quotient. Round answers to the nearest thousandth, if necessary.

1. $5\overline{)34.8}$

    1. _____

2. $-11\overline{)46.98}$

    2. _____

3. $33\overline{)77.847}$

    3. _____

Objective 2   Divide a number by a decimal.

For extra help, see Example 4 on page 359 of your text and Section Lecture video for Section 5.5 and Exercise Solutions Clip 23.

Find each quotient. Round answers to the nearest thousandth, if necessary.

4. $0.9\overline{)3.4166}$

4. _____

5. $3.4\overline{)436.05}$

5. _____

6. $-0.07 \div (-0.00043)$

6. _____

Objective 3   Estimate the answer when dividing decimals.

For extra help, see Example 5 on page 360 of your text and Section Lecture video for Section 5.5 and Exercise Solutions Clip 29.

*Decide if each answer is **reasonable** or **unreasonable** by rounding the numbers and estimating the answer.*

7. $126.2 \div 11.2 = 11.268$

7. _____

8. $31.5 \div 8.4 = 37.5$

8. _____

9. $8695.15 \div 98.762 = 88.0415$

9. _____

Objective 4   Use the order of operations with decimals.

For extra help, see Example 6 on pages 360–361 of your text and Section Lecture video for Section 5.5 and Exercise Solutions Clip 57.

Use the order of operations to simplify each expression.

10. $(-3.1)^2 - 1.9 + 5.8$

10. _____

11. $58.1 - (17.9 - 15.2) \times 1.8$

11. _____

12. $9.1 - 0.07(2.1 \div 0.042)$

12. _____

Chapter 5 RATIONAL NUMBERS: POSITIVE AND NEGATIVE DECIMALS

5.6   Fractions and Decimals

Learning Objectives
1   Write fractions as equivalent decimals.
2   Compare the size of fractions and decimals.

Key Terms

Use the vocabulary terms listed below to complete each statement in exercises 1−4.

numerator   denominator   mixed number   equivalent

1.   A fraction and a decimal that represent the same portion of a whole are

_____.

2.   The _____ of a fraction is the dividend.

3.   The _____ of a fraction shows the number of equal parts in a
whole.

4.   A _____ consists of a whole number part and a fractional or
decimal part.

Guided Examples

Review these examples for Objective 1:
1.   Write the fraction as a decimal.

    a. $\dfrac{7}{8}$

$\dfrac{7}{8}$ means $7 \div 8$. Write it as $8\overline{)7}$. Write extra

zeros in the dividend so you can continue
dividing until the remainder is zero.

$$
\begin{array}{r}
0.875 \\
8\overline{)7.000} \\
\underline{6\,4} \\
60 \\
\underline{56} \\
40 \\
\underline{40} \\
0
\end{array}
$$

Therefore, $\dfrac{7}{8} = 0.875$

Now Try:
1.   Write the fraction as a decimal.

    a. $\dfrac{1}{16}$

_____

b. $3\frac{3}{16}$

One method is to divide 3 by 16 to get 0.1875 for the fraction part. Then add the whole number part to 0.1875.

$$\frac{3}{16} \rightarrow 16)\overline{3.0000} \rightarrow \begin{array}{r} 3.0000 \\ +\ 0.1875 \\ \hline 3.1875 \end{array}$$

$$\begin{array}{r} 0.1875 \\ 16)\overline{3.0000} \\ \underline{1\ 6} \\ 140 \\ \underline{128} \\ 120 \\ \underline{112} \\ 80 \\ \underline{80} \\ 0 \end{array}$$

So $3\frac{3}{16} = 3.1875$.

A second method is to first write $3\frac{3}{16}$ as an improper fraction and then divide numerator by denominator.

$$3\frac{3}{16} = \frac{51}{16}$$

$$\frac{51}{16} \rightarrow 51 \div 16 \rightarrow 16)\overline{51} \rightarrow 16)\overline{51.0000}$$

$$\begin{array}{r} 3.1875 \\ 16)\overline{51.0000} \\ \underline{48} \\ 30 \\ \underline{16} \\ 140 \\ \underline{128} \\ 120 \\ \underline{112} \\ 80 \\ \underline{80} \\ 0 \end{array}$$

So $3\frac{3}{16} = 3.1875$.

2. Write $\frac{5}{9}$ as a decimal and round to the nearest thousandth.

$\frac{5}{9}$ means $5 \div 9$. To round to thousandths, divide out one more place, to ten-thousandths.

b. $4\frac{3}{8}$

_____

2. Write $\frac{5}{12}$ as a decimal and round to the nearest thousandth.

_____

$$\frac{5}{9} \rightarrow 5 \div 9 \rightarrow 9\overline{)5} \rightarrow 9\overline{)5.0000}$$

$$
\begin{array}{r}
0.5555 \\
9\overline{)5.0000} \\
\underline{45} \\
50 \\
\underline{45} \\
50 \\
\underline{45} \\
50 \\
\underline{45} \\
5
\end{array}
$$

Written as a repeating decimal, $\frac{5}{9} = 0.\overline{5}$.

Rounded to the nearest thousandth, $\frac{5}{9} = 0.556$.

Review these examples for Objective 2:

3. Write in $<$, $>$, or $=$ in the blank between each pair of numbers.

a. 0.735 _____ 0.75

Because 0.735 is to the left of 0.75, use the $<$ symbol.
0.735 is less than 0.75 can be written as
   $0.735 < 0.75$

b. $\frac{3}{5}$ _____ 0.6

$\frac{3}{5}$ and 0.6 are the same point on the number line.
They are equivalent.
   $\frac{3}{5} = 0.6$

c. $0.\overline{3}$ _____ 0.3

0.3 is to the right of $0.\overline{3}$ (which is actually 0.333...), so use the $>$ symbol.
   $0.\overline{3}$ is greater than 0.3 can be written as
   $0.\overline{3} > 0.3$

d. $\frac{14}{18}$ _____ $0.\overline{7}$

Write $\frac{14}{18}$ in lowest terms as $\frac{7}{9}$.
We see that $\frac{7}{9} = 0.\overline{7}$.

Now Try:

3. Write in $<$, $>$, or $=$ in the blank between each pair of numbers.

a. 0.85 _____ 0.7895

_____

b. $\frac{5}{16}$ _____ 0.3125

_____

c. $\frac{3}{7}$ _____ $0.\overline{4}$

_____

d. $\frac{27}{33}$ _____ $0.\overline{81}$

_____

4. Write each group of numbers in order, from least to greatest.

a. 0.58 0.575 0.5816

Write zeros to the right of 0.58 and 0.575, so they also have four decimal places. Then find the least and greatest number of ten-thousandths.

0.58 = 0.5800 = 5800 ten-thousandths middle

0.575 = 0.5750 = 5750 ten-thousandths least

= 0.5816 = 5816 ten-thousandths greatest

From least to greatest, the correct order is
0.575 0.58 0.5816

b. $4\frac{7}{8}$ 4.7 4.82

Write $4\frac{7}{8}$ as $\frac{39}{8}$ and divide to get the decimal form 4.875. Then because 4.875 has three decimal places, write zeros so all the numbers have three decimal places.

$4\frac{7}{8}$ = 4.875 = 4 and 875 thousandths greatest

4.7 = 4.700 = 4 and 700 thousandths least

4.82 = 4.820 = 4 and 820 thousandths middle

From least to greatest, the correct order is
4.7 4.82 $4\frac{7}{8}$

4. Write each group of numbers in order, from least to greatest.

a. 0.3 0.307 0.3057

_____

b. $\frac{2}{9}$, $\frac{3}{13}$, 0.23, $\frac{1}{5}$

_____

Objective 1 Write fractions as equivalent decimals.

For extra help, see Examples 1–2 on pages 368–369 of your text and Section Lecture video for Section 5.6 and Exercise Solutions Clip 9, 17, 19, and 27.

Write each fraction or mixed number as a decimal. Round to the nearest thousandth, if necessary.

1. $\frac{1}{8}$

1. _____

2. $4\frac{1}{9}$

2. _____

3. $19\frac{17}{24}$

3. _____

Name: Date:

Instructor: Section:

Objective 2   Compare the size of fractions and decimals.

For extra help, see Examples 3–4 on pages 370–371 of your text and Section Lecture video for Section 5.6 and Exercise Solutions Clip 57.

Write < or > to make a true statement.

4. $\dfrac{5}{6}$ ____ 0.83   4. _____

Arrange in order from smallest to largest.

5. $\dfrac{3}{11}, \dfrac{1}{3}, 0.29$ 5. _____

6. $1.085, 1\dfrac{5}{11}, 1\dfrac{7}{20}$ 6. _____

Chapter 5 RATIONAL NUMBERS: POSITIVE AND NEGATIVE DECIMALS

5.7   Problem Solving with Statistics: Mean, Median, and Mode

Learning Objectives	
1	Find the mean of a list of numbers.
2	Find a weighted mean.
3	Find the median.
4	Find the mode.

Key Terms

Use the vocabulary terms listed below to complete each statement in exercises 1−4.

mean   weighted mean   median   mode

1. The _____ is the value that occurs most often in a group of values.

2. A mean calculated so that each value is multiplied by its frequency is called a _____.

3. The sum of all the values in a data set divided by the number of values in the data set is called the _____.

4. The middle number in a group of values that are listed from smallest to largest is called the _____.

Guided Examples

Review these examples for Objective 1:

1. Find the mean of 125, 234, 155, 275, and 141.

    Use the formula for finding the mean. Add up all the values and then divide by the number of values.

    $$\text{mean} = \frac{125 + 234 + 155 + 275 + 141}{5}$$

    $$\text{mean} = \frac{930}{5}$$

    $$\text{mean} = 186$$

    The mean is 186.

Now Try:

1. Find the mean of 257, 261, 269, 274, 268, 280, and 295.

    _____

2. Find the mean of 31, 37, 44, 51, 52, 74, 69, and 83.

    Find the mean (rounded to the nearest tenth).

    $$\text{mean} = \frac{31 + 37 + 44 + 51 + 52 + 74 + 69 + 83}{8}$$

    $$\text{mean} = \frac{441}{8}$$

    $$\text{mean} \approx 55.1$$

    The mean is 55.1.

2. Find the mean of 40.1, 32.8, 82.5, 51.2, 88.3, 31.7, 43.7, and 51.2.

    _____

Review these examples for Objective 2:

3. Use the following table to find the weighted mean.

Value	Frequency
13	4
12	2
19	5
15	3
21	1
27	5

    To find the mean, multiply the value by the frequency. Then add the products. Next, add the numbers in the frequency column to find the total number of values.

Value	Frequency	Product
13	4	$(13 \cdot 4) = 52$
12	2	$(12 \cdot 2) = 24$
19	5	$(19 \cdot 5) = 95$
15	3	$(15 \cdot 3) = 45$
21	1	$(21 \cdot 1) = 21$
27	5	$(27 \cdot 5) = 135$
Totals	20	372

    Finally, divide the totals. Round to the nearest tenth.

    $$\text{mean} = \frac{372}{20} = 18.6$$

    The mean is 18.6.

Now Try:

3. Use the following table to find the weighted mean.

Value	Frequency
35	1
36	2
39	5
40	4
42	3
43	5

    _____

4. Find the grade point average for this student. Assume A = 4, B = 3, C = 2, D = 1, F = 0.

Units	Grade
3	C
3	A
4	B
5	B
2	A

Multiply the credits and grades. Find the total.

Units	Grade	Units·Grade
3	C(= 2)	3·2 = 6
3	A(= 4)	3·4 = 12
4	B(= 3)	4·3 = 12
5	B(= 3)	5·3 = 15
2	A(= 4)	2·4 = 8
Totals 17		53

It is common to round grade point averages to the nearest hundredth.

$$\text{GPA} = \frac{53}{17} \approx 3.12$$

4. Find the grade point average for this student. Assume A = 4, B = 3, C = 2, D = 1, F = 0.

Units	Grade
3	A
4	B
2	C
5	C
2	D

_____

Review these examples for Objective 3:

5. Find the median for the list of values.

21, 32, 27, 23, 25, 29, 22

First arrange the numbers in numerical order from least to greatest.

21, 22, 23, 25, 27, 29, 32

Next, find the middle number in the list.

21, 22, 23,   25,   27, 29, 32

Three are below.   ↓   Three are above.

Middle number

The median value is 25.

6. Find the median for the list of values.

389, 464, 521, 610, 654, 672, 682, 712

First arrange the numbers in numerical order from least to greatest. Then find the middle two numbers.

389, 464, 521, 610, 654, 672, 682, 712

Middle two numbers

The median value is the mean of the two middle

Now Try:

5. Find the median for the list of values.

18, 12, 11, 19, 26

_____

6. Find the median for the list of values.

0.02, 0.04, 0.12, 0.08

_____

numbers.

$$\text{median} = \frac{610 + 654}{2} = \frac{1264}{2} = 632$$

The median value is 632.

Review these examples for Objective 4:	**Now Try:**
7. Find the mode for each list of numbers.	7. Find the mode for each list of numbers.
a. 37, 52, 41, 27, 41, 96	**a.** 5, 3, 6, 3, 7, 8, 5, 3, 4
The number 41 occurs more often than any other number; therefore, 41 is the mode.	_____
b. 964, 987, 973, 987, 921, 921, 975	**b.** 16, 13, 21, 16, 18, 11, 13, 15, 14
Because both 987 and 921 occur twice, each is a mode.	
	_____
c. $10.71, $11.67, $12.39, $21.54, $13.98, $14.66	**c.** 1.5, 1.2, 1.3, 1.9, 1.8, 1.4, 1.7, 1.0
No number occurs more than once. This list has no mode.	_____

Objective 1   Find the mean of a list of numbers.

For extra help, see Examples 1–2 on page 376 of your text and Section Lecture video for Section 5.7 and Exercise Solutions Clip 3.

Find the mean for each list of numbers. Round to the nearest tenth, if necessary.

1. 39, 50, 59, 61, 69, 73, 51, 80   1. _____

2. 62.7, 59.6, 71.2, 65.8, 63.1   2. _____

3. 216, 245, 268, 268, 280, 291, 304, 313   3. _____

Objective 2   Find a weighted mean.

For extra help, see Examples 3–4 on pages 377–378 of your text and Section Lecture video for Section 5.7 and Exercise Solutions Clip 7.

Find the weighted mean for each list of numbers. Round to the nearest tenth, if necessary.

4.   4. _____

Value	Frequency
17	4
12	5
15	3
19	1

5.

Value	Frequency
1	2
2	3
4	5
5	7
6	4
7	2
8	1
9	1

5. _____

Find the grade point average for this student. Assume A = 4, B = 3, C = 2, D = 1, F = 0.

6.

Units	Grade
5	B
4	C
3	B
2	C
2	C

6. _____

Objective 3   Find the median.

For extra help, see Examples 5–6 on pages 378–379 of your text and Section Lecture video for Section 5.7 and Exercise Solutions Clip 15.

Find the median for each list of numbers.

7.   200, 215, 226, 238, 250, 283

7. _____

8.   43, 69, 108, 32, 51, 49, 83, 57, 64

8. _____

9.   200, 195, 302, 284, 256, 237, 239, 240

9. _____

Objective 4   Find the mode.

For extra help, see Example 7 on page 379 of your text and Section Lecture video for Section 5.7 and Exercise Solutions Clip 25.

Find the mode for each list of numbers.

10.   4, 9, 3, 4, 7, 3, 2, 3, 9

10. _____

11.   37, 24, 35, 35, 24, 38, 39, 28, 27, 39

11. _____

12.   172.6, 199.7, 182.4, 167.1, 172.6, 183.4, 187.6

12. _____

Chapter 5 RATIONAL NUMBERS: POSITIVE AND NEGATIVE DECIMALS

5.8   Geometry Applications: Pythagorean Theorem and Square Roots

Learning Objectives
1   Find square roots using the square root key on a calculator.
2   Find the unknown length in a right triangle.
3   Solve application problems involving right triangles.

Key Terms

Use the vocabulary terms listed below to complete each statement in exercises 1–3.

       hypotenuse   **legs**   **right triangle**

1.   A triangle with a 90° angle is called a _____.

2.   The side opposite the right angle in a right triangle is called the _____ of the triangle.

3.   The two sides of the right angle in a right triangle are called the _____ of the triangle.

Guided Examples

Review these examples for Objective 1:

1.   Use a calculator to find each square root. Round answers to the nearest hundredth.

     a. $\sqrt{45}$

Calculator shows 6.708203932; round to 6.71.

     b. $\sqrt{86}$

Calculator shows 9.273618495; round to 9.27.

     c. $\sqrt{120}$

Calculator shows 10.95445115; round to 10.95.

Now Try:

1.   Use a calculator to find each square root. Round answers to the nearest hundredth.

     a. $\sqrt{20}$

                  _____

     b. $\sqrt{92}$

                  _____

     c. $\sqrt{134}$

                  _____

Name: Date:

Instructor: Section:

Review these examples for Objective 2:

2. Find the unknown length in each right triangle. Round answers to the nearest tenth if necessary.

a.

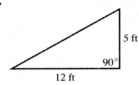

The unknown length is the side opposite the right angle, which is the hypotenuse. Use the formula for finding the hypotenuse.

$$\text{hypotenuse} = \sqrt{(\text{leg})^2 + (\text{leg})^2}$$

$$\text{hypotenuse} = \sqrt{(5)^2 + (12)^2}$$

$$= \sqrt{25 + 144}$$

$$= \sqrt{169}$$

$$= 13$$

The hypotenuse is 13 ft long.

b.

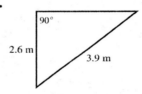

We do know the length of the hypotenuse (3.9 m), so it is the length of one of the legs that is unknown. Use the formula for finding the leg.

$$\text{leg} = \sqrt{(\text{hypotenuse})^2 - (\text{leg})^2}$$

$$\text{leg} = \sqrt{(3.9)^2 - (2.6)^2}$$

$$= \sqrt{15.21 - 6.76}$$

$$= \sqrt{8.45}$$

$$\approx 2.9$$

The length of the leg is approximately 2.9 m.

Now Try:

2. Find the unknown length in each right triangle. Round answers to the nearest tenth if necessary.

a.

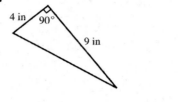

_____

b.

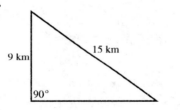

_____

Review this example for Objective 3:

3. The base of a ladder is located 7 feet from a building. The ladder reaches 24 feet up the building. How long is the ladder? Round to the nearest tenth of a foot if necessary.

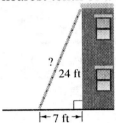

A right triangle is formed. The unknown side is the hypotenuse.

$$\text{hypotenuse} = \sqrt{(\text{leg})^2 + (\text{leg})^2}$$

$$\text{hypotenuse} = \sqrt{(7)^2 + (24)^2}$$

$$= \sqrt{49 + 576}$$

$$= \sqrt{625}$$

$$= 25$$

The length of the ladder is 25 ft.

Now Try:

3. Find the unknown length in this roof plan. Round to the nearest tenth of a foot if necessary.

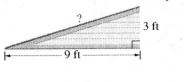

1. _____

Objective 1   Find square roots using the square root key on a calculator.

For extra help, see Example 1 on page 384 of your text and Section Lecture video for Section 5.8 and Exercise Solutions Clip 9.

Find each square root. Use a calculator with a square root key. Round the answer to the nearest thousandth, if necessary.

1. $\sqrt{17}$

2. $\sqrt{75}$

3. $\sqrt{102}$

1. _____

2. _____

3. _____

Name: _____ Date: _____

Instructor: _____ Section: _____

Objective 2   Find the unknown length in a right triangle.

For extra help, see Example 2 on page 385 of your text and Section Lecture video for Section 5.8 and Exercise Solutions Clip 19 and 23.

Find the unknown length in each right triangle. Use a calculator with a square root key. Round the answer to the nearest tenth, if necessary.

4.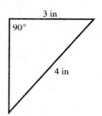

4. _____

5.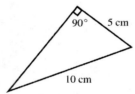

5. _____

6.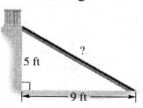

6. _____

Objective 3   Solve application problems involving right triangles.

For extra help, see Example 3 on page 386 of your text and Section Lecture video for Section 5.8 and Exercise Solutions Clip 43.

Solve each application problem. Draw a diagram if one is not provided. Use a calculator with a square root key. Round the answer to the nearest tenth, if necessary.

7.   Find the length of the loading ramp.

7. _____

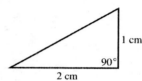

Name: Date:
Instructor: Section:

8.   A kite is flying on 50 feet of string. If the horizontal   8.
      distance of the kite from the person flying it is 40
      feet, how far off the ground is the kite?

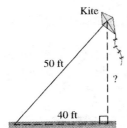

9.   The base of a 17-ft ladder is located 15 ft from a   9.
      building. How high up on the building will the
      ladder reach?

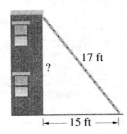

Chapter 5 RATIONAL NUMBERS: POSITIVE AND NEGATIVE DECIMALS

5.9   Problem Solving: Equations Containing Decimals

Learning Objectives	
1	Solve equations containing decimals using the addition property of equality.
2	Solve equations containing decimals using the division property of equality.
3	Solve equations containing decimals using both properties of equality.
4	Solve application problems involving equations with decimals.

Key Terms

Use the vocabulary terms listed below to complete each statement in exercises 1−2.

> **addition property of equality**

> **division property of equality**

1. The _____ states that both sides of an equation may be divided by the same nonzero number and it will still be balanced.

2. The _____ states that the same number may be added to both sides of an equation and the equation will still be balanced.

Guided Examples

Review these examples for Objective 1:

1. Solve each equation and check each solution.

   a. $w + 3.8 = -0.7$

Use the addition property to "get rid of" the 3.8 on the left side by adding its opposite, -3.8.

$$w + 3.8 = -0.7$$
$$\underline{\quad -3.8 \quad -3.8 \quad}$$
$$w + 0 = -4.5$$
$$w \quad = -4.5$$

The solution is -4.5.

Check. Go back to the original equation and replace w with -4.5.

$$w + 3.8 = -0.7$$
$$-4.5 + 3.8 = -0.7$$
$$-0.7 = -0.7$$

So -4.5 is the correct solution.

Now Try:

1. Solve each equation and check each solution.

   a. $w + 4.2 = -9.7$

_____

b. $8 = -5.6 + x$

To get x by itself on the right side of the equal sign, add 5.6 to both sides.

$$8 \quad = -5.6 + x$$

$$\underline{+5.6 \quad +5.6}$$

$$13.6 = \quad 0 + x$$

$$13.6 = \quad\quad x$$

The solution is 13.6.

Check. Go back to the original equation and replace x with 13.6.

$$8 = -5.6 + x$$

$$8 = -5.6 + 13.6$$

$$8 = \quad 8$$

So 13.6 is the correct solution.

b. $11 = -8.6 + x$

_____

Review these examples for Objective 2:

2. Solve each equation and check each solution.

a. $6x = 25.5$

To undo the multiplication by 6, divide both sides by 6.

$$6x = 25.5$$

$$\frac{6 \cdot x}{6} = \frac{25.5}{6}$$

$$\frac{\overset{1}{\cancel{6}} \cdot x}{\underset{1}{\cancel{6}}} = 4.25$$

$$x = 4.25$$

The solution is 4.25.

Check. Go back to the original equation and replace x with 4.25.

$$6x = 25.5$$

$$6(4.25) = 25.5$$

$$25.5 = 25.5$$

So 4.25 is the correct solution.

b. $-10.5 = 1.4t$

Divide both sides by the coefficient of the variable term, 1.4.

Now Try:

2. Solve each equation and check each solution.

a. $8x = 24.4$

_____

b. $-54.4 = 6.4t$

_____

$$-10.5 = 1.4t$$

$$\frac{-10.5}{1.4} = \frac{\overset{1}{\cancel{1.4}}\, t}{\cancel{1.4}}$$

$$-7.5 = t$$

The solution is -7.5.

Check. Go back to the original equation and replace t with -7.5.

$$-10.5 = 1.4t$$

$$-10.5 = 1.4(-7.5)$$

$$-10.5 = -10.5$$

So -7.5 is the correct solution.

Review these examples for Objective 3:	Now Try:
3. Solve each equation and check each solution.	**3.** Solve each equation and check each solution.

a. $3.5b + 0.45 = -6.55$

The first step is to get the variable term, $3.5b$, by itself on the left side of the equal sign.

$$3.5b + 0.45 = -6.55$$
$$\underline{\quad -0.45 \quad -0.45}$$
$$3.5b + 0 = -7.00$$
$$3.5b = -7$$

The next step is to divide both sides by the coefficient of the variable term.

$$\frac{\overset{1}{\cancel{3.5}}\, b}{\cancel{3.5}} = \frac{-7}{3.5}$$

$$b = -2$$

The solution is -2.

Check. Go back to the original equation and replace b with -2.

$$3.5b + 0.45 = -6.55$$
$$3.5(-2) + 0.45 = -6.55$$
$$-7 + 0.45 = -6.55$$
$$-6.55 = -6.55$$

So -2 is the correct solution.

b. $7x - 0.87 = 3x + 2.49$

Use the addition property to "get rid of" $3x$ on the right side by adding its opposite, $-3x$, to both sides.

Now Try column:

3. Solve each equation and check each solution.
a. $4.2b - 0.79 = -7.51$

_____

b. $9x - 1.46 = 3x + 1.72$

_____

$$7x - 0.87 = 3x + 2.49$$
$$\underline{-3x \qquad\qquad -3x}$$
$$4x - 0.87 = 0 \; + 2.49$$
$$4x + (-0.87) = 2.49$$
$$\underline{+0.87 \qquad +0.87}$$
$$4x + \; 0 \; = 3.36$$

$$\frac{\cancel{4}^{1} x}{\cancel{4}_{1}} = \frac{3.36}{4}$$

$$x = 0.84$$

The solution is 0.84.

Check. Go back to the original equation and replace x with 0.84.

$$7x - 0.87 = 3x + 2.49$$
$$7(0.84) - 0.87 = 3(0.84) + 2.49$$
$$5.88 \; - 0.87 = 2.52 + 2.49$$
$$5.01 = 5.01$$

So 0.84 is the correct solution.

Review this example for Objective 4:

4. When using a particular calling card, the cost is $1.15 per minute plus a $0.75 service charge per call. If Julie was billed $20.30 for one call, how long did the call last?

Step 1 Read the problem. It is about the cost of the call.

Step 2 Assign a variable. Let m be the number of minutes for the call.

Step 3 Write an equation.

Cost per minute		Number of minutes		Service charge		Total cost
1.15	$\cdot$	m	$+$	0.75	$=$	20.30

Step 4 Solve the equation.
$$1.15m + 0.75 = 20.30$$
$$\underline{-0.75 \qquad -0.75}$$
$$1.15m + 0 \; = 19.55$$
$$\frac{1.15m}{1.15} = \frac{19.55}{1.15}$$
$$m = 17$$

Now Try:

4. A telephone company charges $4.25 plus an additional $0.07 per minute for long distance calls. How many minutes did Sheila talk if her bill for a call was $6.49?

_____

Step 5 State the answer. The call was 17 min.

Step 6 Check the solution.
   $1.15 per minute times 17 minutes is $19.55
   $19.55 plus $0.75 service charge is $20.30.
The answer checks.

Objective 1   Solve equations containing decimals using the addition property of equality.

For extra help, see Example 1 on page 391 of your text and Section Lecture video for Section 5.9.

Solve each equation and check each solution.

1. $-30.4 + n = -35$ 1. _____

2. $-7.1 = 0.32 + m$ 2. _____

3. $8.7 + h = 16.4$ 3. _____

Objective 2   Solve equations containing decimals using the division property of equality.

For extra help, see Example 2 on page 392 of your text and Section Lecture video for Section 5.9 and Exercise Solutions Clip 7 and 11.

Solve each equation and check each solution.

4. $-3y = -0.96$ 4. _____

5. $2.7r = 6.75$ 5. _____

6. $87.6 = -12r$ 6. _____

Objective 3   Solve equations containing decimals using both properties of equality.

For extra help, see Example 3 on page 393 of your text and Section Lecture video for Section 5.9 and Exercise Solutions Clip 19.

Solve each equation and check each solution.

7. $2.2 = 0.5y + 9.7$ 7. _____

8. $-11.2 - 0.7p = 0.9p + 5.6$ 8. _____

9. $0.6x + 4.98 = x - 6.78$ 9. _____

Objective 4   Solve application problems involving equations with decimals.

For extra help, see Example 4 on page 394 of your text and Section Lecture video for Section 5.9 and Exercise Solutions Clip 25.

Solve each application problem using the six problem-solving steps.

10. An air compressor can be rented for $38.95 for the   10. _____
first three hours, and $8.50 for each additional hour.
William's rental charge was $64.45. How many
hours did he rent the compressor?

11. Most adult medication doses are for a person   11. _____
weighing 150 pounds. For a 45-pound child, the
adult dose should be multiplied by 0.3. If the child's
dose of a decongestant is 9 milligrams, what is the
adult dose?

12. The bill for a three-night hotel stay was $371.43.   12. _____
This included $42.15 for room service and $35.28
room tax. What was the room rate per night?

Chapter 5 RATIONAL NUMBERS: POSITIVE AND NEGATIVE DECIMALS

5.10   Geometry Applications: Circles, Cylinders, and Surface Area

Learning Objectives	
1	Find the radius and diameter of a circle.
2	Find the circumference of a circle.
3	Find the area of a circle.
4	Find the volume of a cylinder.
5	Find the surface area of a rectangular solid.
6	Find the surface area of a cylinder.

Key Terms

Use the vocabulary terms listed below to complete each statement in exercises 1−6.

    circle   radius   diameter   circumference   π (pi)

    surface area

1.   The _____ is the distance from the center of a circle to any point
      on the circle.

2.   The _____ of a circle is the distance around the circle.

3.   A figure whose points lie the same distance from a fixed center point is called a
      _____.

4.   The ratio of the circumference to the diameter of any circle equals _____.

5.   The area on the surface of a three-dimensional object is called its _____.

6.   The _____ of a circle is a segment connecting two points
      on a circle and passing through the center.

Guided Examples

Review these examples for Objective 1:
1.   Find the unknown length of the diameter or
      radius of each circle.

      a.

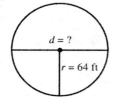

      Because the radius is 64 ft, the diameter is twice

Now Try:
1.   Find the unknown length of the
      diameter or radius of each circle.

      a.

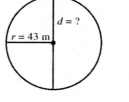

      _____

Copyright © 2014 Pearson Education, Inc.

as long.

$$d = 2 \cdot r$$

$$d = 2 \cdot 64 \text{ ft}$$

$$d = 128 \text{ ft}$$

b.

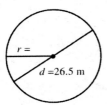

b.

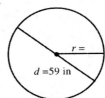

_____

The radius is half the diameter.

$$r = \frac{d}{2}$$

$$r = \frac{26.5 \text{ m}}{2}$$

$$r = 13.25 \text{ m}$$

Review these examples for Objective 2:

2. Find the circumference of each circle. Use 3.14 as the approximate value for π. Round answers to the nearest tenth.

a.

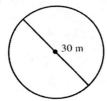

The diameter is 30 m, so use the formula with d in it.

$$C = \pi \cdot d$$

$$C \approx 3.14 \cdot 30 \text{ m}$$

$$C \approx 94.2 \text{ m}$$

b.

In this example, the radius is labeled, so it is easier to use the formula with r in it.

Now Try:

2. Find the circumference of each circle. Use 3.14 as the approximate value for π. Round answers to the nearest tenth.

a.

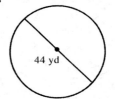

_____

b.

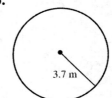

_____

$$C = 2 \cdot \pi \cdot r$$

$$C \approx 2 \cdot 3.14 \cdot 23 \text{ cm}$$

$$C \approx 144.4 \text{ cm} \quad \text{Rounded}$$

Review these examples for Objective 3:	**Now Try:**
3. Find the area of each circle. Use 3.14 as the approximate value for π. Round your answers to the nearest tenth.	3. Find the area of each circle. Use 3.14 as the approximate value for π. Round your answers to the nearest tenth.

a.

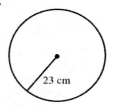

Use the formula $A = \pi \cdot r^2$, which means $\pi \cdot r \cdot r$.

$$A = \pi \cdot r \cdot r$$

$$A \approx 3.14 \cdot 23 \text{ cm} \cdot 23 \text{ cm}$$

$$A \approx 1661.1 \text{ cm}^2 \quad \text{Rounded}$$

b.

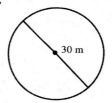

To use the area formula, you need to know the radius (r). In this circle, the diameter is 30 m. First find the radius.

$$r = \frac{d}{2}$$

$$r = \frac{30 \text{ m}}{2} = 15 \text{ m}$$

Now find the area.

$$A = \pi \cdot r \cdot r$$

$$A \approx 3.14 \cdot 15 \text{ m} \cdot 15 \text{ m}$$

$$A \approx 706.5 \text{ m}^2$$

a.

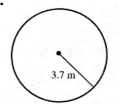

_____

b.

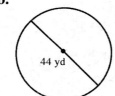

_____

4. Find the area of a semicircle with radius 13 cm. Use 3.14 as the approximate value for π. Round your answers to the nearest tenth.

First, find the area of a whole circle with a radius of 13 cm.

$$A = \pi \cdot r \cdot r$$

$$A \approx 3.14 \cdot 13 \text{ m} \cdot 13 \text{ m}$$

$$A \approx 530.66 \text{ m}^2$$

Divide the area of the whole circle by 2 to find the area of the semicircle.

$$\frac{530.66 \text{ m}^2}{2} = 265.33 \text{ m}^2$$

The last step is rounding 265.33 to the nearest tenth.

Area of semicircle $\approx 265.3 \text{ m}^2$ Rounded

4. Find the area of a semicircle with radius 9 ft. Use 3.14 as the approximate value for π. Round your answers to the nearest tenth.

_____

5. A circular piece of glass, for a customized table, is 3 ft in diameter. The cost of metal for the edge is $3.50 per foot. What will it cost to add a metal edge to the glass? Use 3.14 for π.

$$\text{Circumference} = \pi \cdot d$$

$$C \approx 3.14 \cdot 3 \text{ ft}$$

$$C \approx 9.42 \text{ ft}$$

$$\text{cost} = \text{cost per foot} \cdot \text{circumference}$$

$$\text{cost} = \frac{\$3.50}{1 \text{ ft}} \cdot \frac{9.42 \text{ ft}}{1}$$

$$\text{cost} = \$32.97$$

The cost of adding a metal edge to the glass is $32.97.

5. A circular coaster is 5 inches in diameter. The cost of rubber for the edge is $0.80 per inch. What will it cost to add rubber edge to the coaster? Use 3.14 for π.

_____

6. Find the cost of covering the glass in Example 5 with a plastic cover. The material for the cover costs $2.50 per square foot. Use 3.14 for π. Round your answer to the nearest cent.

First find the radius. $r = \dfrac{d}{2} = \dfrac{3 \text{ ft}}{2} = 1.5 \text{ ft}$

Then find the area. $A = \pi \cdot r^2$

$$A \approx 3.14 \cdot 1.5 \text{ ft} \cdot 1.5 \text{ ft}$$

$$A \approx 7.065 \text{ ft}^2$$

6. Find the cost of covering the coaster in Example 5 with a protective coating. The material for the cover costs $0.10 per square inch. Use 3.14 for π. Round your answer to the nearest cent.

_____

187

$$\text{cost} = \frac{\$2.50}{1 \ \cancel{ft^2}} \cdot \frac{7.065 \ \cancel{ft^2}}{1}$$

$$\text{cost} = \$17.66 \quad \text{Rounded}$$

The cost of the plastic cover is $17.66.

Review these examples for Objective 4:

7. Find the volume of each cylinder. Use 3.14 as the approximate value of π. Round your answers to the nearest tenth if necessary.

a.

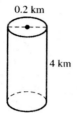

0.2 km

4 km

The diameter is 0.2 km, so the radius is 0.2 km $\div 2 = 0.1$ km. The height is 4 km.

$$V = \pi \cdot r \cdot r \cdot h$$

$$V \approx 3.14 \cdot 0.1 \text{ km} \cdot 0.1 \text{ km} \cdot 4 \text{ km}$$

$$V \approx 0.1 \text{ km}^3$$

b.

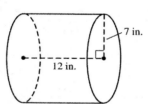

7 in.

12 in.

$$V = \pi \cdot r \cdot r \cdot h$$

$$V \approx 3.14 \cdot 7 \text{ in.} \cdot 7 \text{ in.} \cdot 12 \text{ in.}$$

$$V \approx 1846.3 \text{ in.}^3$$

Now Try:

7. Find the volume of each cylinder. Use 3.14 as the approximate value of π. Round your answers to the nearest tenth if necessary.
    a. A cardboard mailing tube, diameter 5 centimeters and height 25 centimeters

    _____

    b. An oil can, diameter 8 centimeters and height 13.5 centimeters

    _____

Review this example for Objective 5:

8. Find the volume and surface area of the figure.

6.8 cm

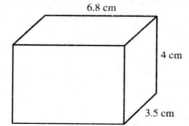

4 cm

3.5 cm

First find the volume.

Now Try:

8. Find the volume and surface area of the figure.

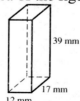

39 mm

17 mm

12 mm

_____

$V = lwh$

$V = 6.8 \text{ cm} \cdot 3.5 \text{ cm} \cdot 4 \text{ cm}$

$V = 95.2 \text{ cm}^3$

Now find the surface area.

$S = 2lw + 2lh + 2wh$

$S = (2 \cdot 6.8 \text{ cm} \cdot 3.5 \text{ cm}) + (2 \cdot 6.8 \text{ cm} \cdot 4 \text{ cm})$
$\quad + (2 \cdot 3.5 \text{ cm} \cdot 4 \text{ cm})$

$S = 47.6 \text{ cm}^2 + 54.4 \text{ cm}^2 + 28 \text{ cm}^2$

$S = 130 \text{ cm}^2$

Review this example for Objective 6:

9. Find the volume and surface area of the figure. Use 3.14 as the approximate value of π. Round your answers to the nearest tenth if necessary.

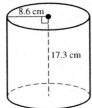

First find the volume.

$V = \pi r^2 h$

$V \approx 3.14 \cdot 8.6 \text{ cm} \cdot 8.6 \text{ cm} \cdot 17.3 \text{ cm}$

$V \approx 4017.7 \text{ cm}^3$

Now find the surface area.

$S = 2\pi rh + 2\pi r^2$

$S \approx (2 \cdot 3.14 \cdot 8.6 \text{ cm} \cdot 17.3 \text{ cm})$
$\quad + (2 \cdot 3.14 \cdot 8.6 \text{ cm} \cdot 8.6 \text{ cm})$

$S \approx 934.3384 \text{ cm}^2 + 464.4688 \text{ cm}^2$

$S \approx 1398.8 \text{ cm}^2$

Now Try:

9. Find the volume and surface area of the figure. Use 3.14 as the approximate value of π. Round your answers to the nearest tenth if necessary.

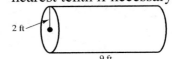

_____

Objective 1   Find the radius and diameter of a circle.

For extra help, see Example 1 on pages 397–398 of your text and Section Lecture video for Section 5.10 and Exercise Solutions Clip 5.

Find the diameter or radius in each circle.

1. The diameter of a circle is 8 feet. Find its radius.

    1. _____

2. The radius of a circle is 2.7 centimeters. Find its diameter.

    2. _____

3. The diameter of a circle is $12\frac{1}{2}$ yards. Find its radius.

3. _____

Objective 2   Find the circumference of a circle.

For extra help, see Example 2 on page 399 of your text and Section Lecture video for Section 5.10 and Exercise Solutions Clip 13.

Find the circumference of each circle. Use 3.14 as an approximation for π. Round each answer to the nearest tenth.

4. A circle with a diameter of $4\frac{3}{4}$ inches

4. _____

5. A circle with a radius of 4.5 yards

5. _____

6. A circle with a radius of 8 cm

6. _____

Objective 3   Find the area of a circle.

For extra help, see Examples 3–6 on pages 400–402 of your text and Section Lecture video for Section 5.10 and Exercise Solutions Clip 13.

Find the area of each circle. Use 3.14 as an approximation for π. Round each answer to the nearest tenth.

7. A circle with diameter of $5\frac{1}{3}$ yards

7. _____

8. A circle with radius of 9.8 centimeters

8. _____

9.

9. _____

Name: Date:

Instructor: Section:

Objective 4   Find the volume of a cylinder.

For extra help, see Example 7 on page 402 of your text and Section Lecture video for Section 5.10 and Exercise Solutions Clip 35 and 39.

Find the volume of each figure. Use 3.14 as an approximation for π. *Round answers to the nearest tenth, if necessary.*

10.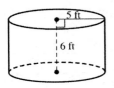

10. _____

11.   18 cm
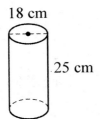

11. _____

12.   A soup can, diameter 6 inches and height 8 inches

12. _____

Objective 5   Find the surface area of a rectangular solid.

For extra help, see Example 8 on page 403 of your text and Section Lecture video for Section 5.10 and Exercise Solutions Clip 47.

Find the surface area of each rectangular solid. Round your answers to the nearest tenth.

13.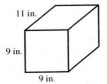

13. _____

14.

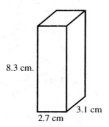

8.3 cm.

3.1 cm

2.7 cm

14. _____

15.

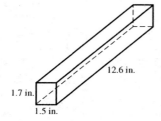

12.6 in.

1.7 in.

1.5 in.

15. _____

Objective 6   Find the surface area of a cylinder.

For extra help, see Example 9 on page 404 of your text and Section Lecture video for Section 5.10 and Exercise Solutions Clip 35 and 39.

Find the surface area of each cylinder. Use 3.14 as the approximate value for π. Round your answers to the nearest tenth.

16.   24 in.

8 in.

16. _____

17.

3.7 ft

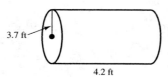

4.2 ft

17. _____

18.   24 in.

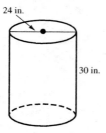

30 in.

18. _____

Chapter 6 RATIO, PROPORTION, AND LINE/ANGLE/TRIANGLE RELATIONSHIPS

6.1   Ratios

Learning Objectives
1   Write ratios as fractions.
2   Solve ratio problems involving decimals or mixed numbers.
3   Solve ratio problems after converting units.

Key Terms

Use the vocabulary terms listed below to complete each statement in exercises 1–2.

denominator   **numerator**   **ratio**

1. A _____ can be used to compare two measurements with the same type of units.

2. When writing the ratio to compare the width of a room to its height, the width

    goes in the _____ and the height goes in the

    _____.

Guided Examples

Review these examples for Objective 1:

1. A meeting room measures 22 feet long, 17 feet wide and 13 feet high. Write each ratio as a fraction, using the room measurements.

    a. Ratio of length to width

    The ratio of length to width is $\dfrac{22 \ \cancel{ft}}{17 \ \cancel{ft}} = \dfrac{22}{17}$

    b. Ratio of width to height.

    The ratio of width to height is $\dfrac{17 \ \cancel{ft}}{13 \ \cancel{ft}} = \dfrac{17}{13}$

Now Try:

1. Jacob spent $5 for breakfast, $8 for lunch, and $17 for dinner. Write each ratio as a fraction.

    a. Ratio of the amount spent on a dinner to the amount spent on lunch

    _____

    b. Ratio of the amount spent on lunch to the amount spent on breakfast

    _____

2. Write each ratio in lowest terms.

a. 32 inches of snow to 8 inches of rain

The ratio is $\dfrac{32 \text{ inches}}{8 \text{ inches}}$. Divide out the common units. Then write this ratio in lowest terms by dividing the numerator and denominator by 8.

$$\frac{32 \ \cancel{\text{inches}}}{8 \ \cancel{\text{inches}}} = \frac{32}{8} = \frac{32 \div 8}{8 \div 8} = \frac{4}{1}$$

So, the ratio of 32 inches of snow to 8 inches of rain is 4 to 1, or $\dfrac{4}{1}$. For every 4 inches of snow there is 1 inch of rain.

b. 12 ounces of cleaner to 40 ounces of water

The ratio is $\dfrac{12 \text{ ounces}}{40 \text{ ounces}}$. Divide out the common units. Then write this ratio in lowest terms by dividing the numerator and denominator by 4.

$$\frac{12 \ \cancel{\text{ounces}}}{40 \ \cancel{\text{ounces}}} = \frac{12}{40} = \frac{12 \div 4}{40 \div 4} = \frac{3}{10}$$

So, the ratio of 12 ounces of cleaner to 40 ounces of water is $\dfrac{3}{10}$. For every 3 ounces of cleaner, there are 10 ounces of water.

c. 14 people at a large table to 8 people at a small table

The ratio is $\dfrac{14}{8} = \dfrac{14 \div 2}{8 \div 2} = \dfrac{7}{4}$

Review these examples for Objective 2:

3. The price of a newspaper increased from $0.75 to $1.00. Find the ratio of the increase in price to the original price.

new price − original price = increase

$1.00 − $0.75 = $0.25

The words "the original price" are mentioned second, so the original price of $0.75 is in the denominator.
The ratio of increase in price to original price is shown below.

$$\frac{0.25}{0.75}$$

Now rewrite the ratio as a ratio of whole

2. Write each ratio in lowest terms.

a. 16 seconds to 24 seconds

_____

b. 150 grams to 75 grams

_____

c. width of 14 m to length of 21 m

_____

Now Try:

3. The original price of a burger is $4.80 and the sale price is $3.00. Find the ratio of the decrease in price to the original price.

_____

numbers. Start by multiplying both the numerator and denominator by 100.

$$\frac{0.25}{0.75} = \frac{0.25 \times 100}{0.75 \times 100} = \frac{25}{75} = \frac{25 \div 25}{75 \div 25} = \frac{1}{3}$$

4. Write each ratio as a comparison of whole numbers in lowest terms.

   a. 5 hours to $5\frac{1}{3}$ hours

   Divide out the common units.

   $$\frac{5 \ \cancel{\text{hours}}}{5\frac{1}{3} \ \cancel{\text{hours}}} = \frac{5}{5\frac{1}{3}}$$

   Next, write 5 as $\frac{5}{1}$ and $5\frac{1}{3}$ as the improper fraction $\frac{16}{3}$.

   $$\frac{5}{5\frac{1}{3}} = \frac{\frac{5}{1}}{\frac{16}{3}}$$

   Now, rewrite the problem in horizontal format, using the " $\div$ " symbol for division. Finally, multiply by the reciprocal of the divisor.

   $$\frac{\frac{5}{1}}{\frac{16}{3}} = \frac{5}{1} \div \frac{16}{3} = \frac{5}{1} \cdot \frac{3}{16} = \frac{15}{16}$$

   The ratio, in lowest terms, is $\frac{15}{16}$.

   b. $7\frac{1}{5}$ to $2\frac{1}{4}$

   Write the ratio as $\dfrac{7\frac{1}{5}}{2\frac{1}{4}}$. Then write $7\frac{1}{5}$ and $2\frac{1}{4}$ as improper fractions.

   $$7\frac{1}{5} = \frac{36}{5} \qquad \text{and} \qquad 2\frac{1}{4} = \frac{9}{4}$$

   The ratio is shown here.

   $$\frac{7\frac{1}{5}}{2\frac{1}{4}} = \frac{\frac{36}{5}}{\frac{9}{4}}$$

   Rewrite as a division problem in horizontal

4. Write each ratio as a comparison of whole numbers in lowest terms.

   a. $6\frac{1}{3}$ to 3

   _____

   b. $3\frac{2}{3}$ to $2\frac{5}{6}$

   _____

format, using the " ÷ " symbol for division. Then multiply by the reciprocal of the divisor.

$$\frac{36}{5} \div \frac{9}{4} = \frac{\overset{4}{\cancel{36}}}{5} \cdot \frac{4}{\underset{1}{\cancel{9}}} = \frac{16}{5}$$

Review this example for Objective 3:	**Now Try:**
5. Write the ratio of length of a board 3 ft long to the length of another board that is 42 inches long.	**5.** Write the ratio of 20 days to 4 weeks.

First, express 3 ft in inches. Because 1 ft has 12 in., 3 ft is
   3·12 in. = 36 in.
The ratio of the lengths is

$$\frac{3 \text{ ft}}{42 \text{ in.}} = \frac{36 \text{ in.}}{42 \text{ in.}} = \frac{36}{42}$$

Write the ratio in lowest terms.

$$\frac{36}{42} = \frac{36 \div 6}{42 \div 6} = \frac{6}{7}$$

The shorter board is $\frac{6}{7}$ the length of the longer board.

Objective 1   Write ratios as fractions.

For extra help, see Examples 1–2 on pages 428–430 of your text and Section Lecture video for Section 6.1 and Exercise Solutions Clip 9.

Write each ratio as a fraction in lowest terms.

   1.   125 cents to 95 cents   **1.** _____

   2.   80 miles to 30 miles   **2.** _____

   3.   5 men to 20 men   **3.** _____

Objective 2   Solve ratio problems involving decimals or mixed numbers.

For extra help, see Examples 3–4 on pages 430–431 of your text and Section Lecture video for Section 6.1 and Exercise Solutions Clip 13 and 33.

Write each ratio as a fraction in lowest terms.

   4. $4\frac{1}{8}$ to 3   **4.** _____

5. 11 to $2\frac{4}{9}$ 5. _____

Solve. Write each ratio as a fraction in lowest terms.

6. One car has a $15\frac{1}{2}$ gallon gas tank while another has   6. _____
a 22 gallon gas tank. Find the ratio of the amount the
first tank holds to the amount the second tank holds.

Objective 3   Solve ratio problems after converting units.

For extra help, see Example 5 on page 432 of your text and Section Lecture video for
Section 6.1 and Exercise Solutions Clip 21.

Write each ratio as a fraction in lowest terms. Be sure to convert units as necessary.

7. 4 days to 2 weeks   7. _____

8. 6 yards to 10 feet   8. _____

9. 40 ounces to 3 pounds   9. _____

Chapter 6 RATIO, PROPORTION, AND LINE/ANGLE/TRIANGLE RELATIONSHIPS

6.2   Rates

Learning Objectives
1   Write rates as fractions.
2   Find unit rates.
3   Find the best buy based on cost per unit.

Key Terms

Use the vocabulary terms listed below to complete each statement in exercises 1–3.

      rate   **unit rate**   **cost per unit**

1.   When the denominator of a rate is 1, it is called a _____.

2.   The _____ is that rate that tells how much is paid for one item.

3.   A _____ compares two measurements with different units.

Guided Examples

Review these examples for Objective 1:

1.   Write each rate as a fraction in lowest terms.

    a. 7 gallons for $56

Write the units: gallons and dollars
$$\frac{7 \text{ gallons} \div 7}{56 \text{ dollars} \div 7} = \frac{1 \text{ gallon}}{8 \text{ dollars}}$$

    b. 75 inches of growth in 15 weeks

$$\frac{75 \text{ inches} \div 15}{15 \text{ weeks} \div 15} = \frac{5 \text{ inches}}{1 \text{ week}}$$

    c. 984 miles on 32 gallons of gas

$$\frac{984 \text{ miles} \div 8}{32 \text{ gallons} \div 8} = \frac{123 \text{ miles}}{4 \text{ gallons}}$$

Now Try:

1.   Write each rate as a fraction in lowest terms.

    a. $7 for 35 pages

        _____

    b. 300 strokes in 20 minutes

        _____

    c. 396 strawberries for 24 cakes

        _____

Review these examples for Objective 2:

2. Find each unit rate.

    a. 478.5 miles on 14.5 gallons of gas

    Write the rate as a fraction.

    $$\frac{478.5 \text{ miles}}{14.5 \text{ gallons}}$$

    Divide 478.5 by 14.5 to find the unit rate.

    $$145\overline{)4785}^{33}$$

    $$\frac{478.5 \text{ miles} \div 14.5}{14.5 \text{ gallons} \div 14.5} = \frac{33 \text{ miles}}{1 \text{ gallon}}$$

    The unit rate is 33 miles per gallon
    or 33 miles/gallon.

    b. 770 miles in 22 hours

    $$\frac{770 \text{ miles}}{22 \text{ hours}} \quad \text{Divide: } 22\overline{)770}^{35}$$

    The unit rate is 35 miles/hour.

    c. $1240 in 8 days

    $$\frac{1240 \text{ dollars}}{8 \text{ days}} \quad \text{Divide: } 8\overline{)1240}^{155}$$

    The unit rate is $155/day.

Now Try:

2. Find each unit rate.

    a. 294 miles on 10.5 gallons of gas

    _____

    b. $7.56 for 6 pounds of apples

    _____

    c. $580 in 4 days

    _____

Review these examples for Objective 3:

3. Determine the best price for peanut butter. For 18 ounces the price is $2.89, for 28 ounces the price is $3.99, and for 40 ounces, the price is $6.18.

    For 18 ounces, the cost per ounce is

    $$\frac{\$2.89}{18 \text{ ounces}} \approx \$0.16$$

    For 28 ounces, the cost per ounce is

    $$\frac{\$3.99}{28 \text{ ounces}} \approx \$0.14$$

    For 40 ounces, the cost per ounce is

    $$\frac{\$6.18}{40 \text{ ounces}} \approx \$0.15$$

    The lowest cost per ounce is $0.14, so the 28-ounce container is the best buy.

Now Try:

3. Find the best buy:
    2 pints for $3.55,
    3 pints for $5.25, and
    5 pints for $8.50.

    _____

4. Solve each application problem.

a. Brand AA laundry detergent costs $5.99 for 32 ounces. Brand ZZ laundry detergent costs $13.29 for 100 ounces. Which choice is the best buy?

To find Brand AA's unit cost, divide $5.99 by 32 ounces. Similarly, to find Brand ZZ's unit cost, divide $13.29 by 100 ounces.

Brand AA $\dfrac{\$5.99}{32 \text{ ounces}} \approx 0.187$ per ounce

Brand ZZ $\dfrac{\$13.29}{100 \text{ ounces}} = 0.1329$ per ounce

Brand ZZ has the lower cost per ounce and is the better buy.

b. Special offers affect the best buy. Brand M of multivitamin costs $3.99 for 200 + 100 tablets (a bonus of 100 tablets). Brand N of multivitamin costs $5.47 for 200 tablets with a special Buy one Get one Free. Which choice is the best buy?

Brand M is $3.99 for 300 tablets (200 + 100)
Brand N is $5.47 for 400 tablets (200 + 200)

To find the best buy, divide to find the lowest cost per tablet.

Brand M $\dfrac{\$3.99}{300 \text{ tablets}} \approx \$0.0133$ per tablet

Brand N $\dfrac{\$5.47}{400 \text{ tablets}} \approx \$0.0137$ per tablet

Brand M has the lower cost per tablet and is the best buy.

4. Solve each application problem.

a. An eight-pack of AA-size batteries costs $4.99. A twenty-pack of AA-size batteries costs $12.99. Which battery pack is the best buy?

_____

b. Which shampoo is the best buy? You have a coupon for 75¢ off Brand S and a coupon for $1.25 off Brand T. Brand S is $4.79 for 14 ounces. Brand T is $5.69 for 18 ounces.

_____

Objective 1   Write rates as fractions.

For extra help, see Example 1 on page 435 of your text and Section Lecture video for Section 6.2 and Exercise Solutions Clip 1, 3, and 5.

Write each rate as a fraction in lowest terms.

1.   119 pills for 17 patients

1. _____

2.   28 dresses for 4 women

2. _____

3.   256 pages for 8 chapters

3. _____

Name: Date:

Instructor: Section:

Objective 2   Find unit rates.

For extra help, see Example 2 on page 436 of your text and Section Lecture video for Section 6.2 and Exercise Solutions Clip 9 and 11.

Find each unit rate.

4. $3500 in 20 days   4. _____

5. $7875 for 35 pounds   5. _____

6. 189.88 miles on 9.4 gallons   6. _____

Objective 3   Find the best buy based on cost per unit.

For extra help, see Examples 3–4 on pages 436–438 of your text and Section Lecture video for Section 6.2 and Exercise Solutions Clip 21 and 23.

Find the best buy (based on cost per unit) for each item.

7. Peanut butter: 18 ounces for $1.77;   7. _____
    24 ounces for $2.08

8. Batteries: 4 for $2.79; 10 for $4.19   8. _____

9. Soup: 3 cans for $1.75; 5 cans for $2.75;   9. _____
    8 cans for $4.55

Chapter 6 RATIO, PROPORTION, AND LINE/ANGLE/TRIANGLE RELATIONSHIPS

6.3   Proportions

Learning Objectives
1   Write proportions.
2   Determine whether proportions are true or false.
3   Find the unknown number in a proportion.

Key Terms

Use the vocabulary terms listed below to complete each statement in exercises 1−3.

cross products   **proportion**   **ratio**

1.   A _____ shows that two ratios or rates are equivalent.

2.   To see whether a proportion is true, determine if the _____ are equal.

3.   A _____ is a comparison of two quantities with the same units.

Guided Examples

Review these examples for Objective 1:

1.   Write each proportion.

   a. 7 m is to 13 m as 28 m is to 52 m

   $$\frac{7 \text{ m}}{13 \text{ m}} = \frac{28 \text{ m}}{52 \text{ m}} \quad \text{so} \quad \frac{7}{13} = \frac{28}{52}$$

   b. $14 is to 8 gallons as $7 is to 4 gallons

   $$\frac{\$14}{8 \text{ gallons}} = \frac{\$7}{4 \text{ gallons}}$$

Now Try:

1.   Write each proportion.

   a. 24 ft is to 17 ft as 72 ft is to 51 ft

   _____

   b. $10 is to 7 cans as $60 is to 42 cans

   _____

Review these examples for Objective 2:

2. Determine whether each proportion is true or false by writing both ratios in lowest terms.

    a. $\dfrac{7}{11} = \dfrac{16}{24}$

    Write each ratio in lowest terms.

    $\dfrac{7}{11} \leftarrow$ Already in lowest terms $\dfrac{16 \div 8}{24 \div 8} = \dfrac{2}{3} \leftarrow$ Lowest terms

    Because $\dfrac{7}{11}$ is not equivalent to $\dfrac{2}{3}$, the proportion is false.

    b. $\dfrac{9}{15} = \dfrac{21}{35}$

    Write each ratio in lowest terms.

    $\dfrac{9 \div 3}{15 \div 3} = \dfrac{3}{5}$ $\dfrac{21 \div 7}{35 \div 7} = \dfrac{3}{5}$

    Both ratios are equivalent to $\dfrac{3}{5}$, so the proportion is true.

3. Use cross products to see whether each proportion is true or false.

    a. $\dfrac{5}{8} = \dfrac{30}{48}$

    Multiply along one diagonal, then multiply along the other diagonal.

    $\dfrac{5}{8} = \dfrac{30}{48}$ $\nearrow 8 \cdot 30 = 240$
    $\searrow 5 \cdot 48 = 240$

    The cross products are equal, so the proportion is true.

    b. $\dfrac{3\frac{1}{5}}{4\frac{2}{3}} = \dfrac{7}{10}$

    $\dfrac{3\frac{1}{5}}{4\frac{2}{3}} = \dfrac{7}{10}$

    $\nearrow 4\frac{2}{3} \cdot 7 = \dfrac{14}{3} \cdot \dfrac{7}{1} = \dfrac{98}{3} = 32\frac{2}{3}$

    $\searrow 3\frac{1}{5} \cdot 10 = \dfrac{16}{\cancel{5}} \cdot \dfrac{\overset{2}{\cancel{10}}}{1} = \dfrac{32}{1} = 32$

Now Try:

2. Determine whether each proportion is true or false by writing both ratios in lowest terms.

    a. $\dfrac{36}{28} = \dfrac{24}{18}$

    _____

    b. $\dfrac{4}{12} = \dfrac{9}{27}$

    _____

3. Use cross products to see whether each proportion is true or false.

    a. $\dfrac{6}{17} = \dfrac{18}{51}$

    _____

    b. $\dfrac{3.2}{5} = \dfrac{7}{10}$

    _____

The cross products are unequal, so the proportion is false.

Review these examples for Objective 3:	**Now Try:**
4. Find the unknown number in each proportion. Round answers to the nearest hundredth when necessary.	4. Find the unknown number in each proportion. Round answers to the nearest hundredth when necessary.

Review these examples for Objective 3:

4. Find the unknown number in each proportion. Round answers to the nearest hundredth when necessary.

a. $\dfrac{14}{x} = \dfrac{21}{18}$

Recall that ratios can be rewritten in lowest terms. Write $\dfrac{21}{18}$ in lowest terms as $\dfrac{7}{6}$, which gives the proportion $\dfrac{14}{x} = \dfrac{7}{6}$.

Step 1 Find the cross product.

$$\dfrac{14}{x} = \dfrac{7}{6} \quad \nearrow x \cdot 7 \\ \searrow 14 \cdot 6$$

Step 2 Show that cross products are equal.

$x \cdot 7 = 14 \cdot 6$

$x \cdot 7 = 84$

Step 3 Divide both sides by 7.

$$\dfrac{x \cdot \cancel{7}}{\cancel{7}} = \dfrac{84}{7}$$

$x = 12$

Step 4 Check in original proportion.

$$\dfrac{14}{12} = \dfrac{21}{18} \quad \nearrow 12 \cdot 21 = 252 \\ \searrow 14 \cdot 18 = 252$$

The cross products are equal, so 12 is the correct solution.

b. $\dfrac{9}{13} = \dfrac{21}{x}$

Step 1 Find the cross product.

$$\dfrac{9}{13} = \dfrac{21}{x} \quad \nearrow 13 \cdot 21 = 273 \\ \searrow 9 \cdot x$$

Step 2 Show that cross products are equal.

$9 \cdot x = 273$

Step 3 Divide both sides by 9.

Now Try:

4. Find the unknown number in each proportion. Round answers to the nearest hundredth when necessary.

a. $\dfrac{24}{x} = \dfrac{9}{12}$

_____

b. $\dfrac{2}{3} = \dfrac{x}{16}$

_____

$$\frac{\cancel{9}^{\,1} \cdot x}{\cancel{9}_{\,1}} = \frac{273}{9}$$

$$x = 30.33 \text{ rounded to the nearest hundredth}$$

Step 4 Check in original proportion.

$$\frac{9}{13} = \frac{21}{30.33} \quad \begin{array}{l} \nearrow 13 \cdot 21 = 273 \\ \searrow 9 \cdot 30.33 = 272.97 \end{array}$$

The cross products are slightly different because of the rounded value of x. However, they are close enough to see that the problem was done correctly and that 30.33 is the approximate solution.

5. Find the unknown number in each proportion.

a. $\dfrac{3\frac{1}{4}}{8} = \dfrac{x}{12}$

$$\frac{3\frac{1}{4}}{8} = \frac{x}{12} \quad \begin{array}{l} \nearrow 8 \cdot x \\ \searrow 3\frac{1}{4} \cdot 12 \end{array}$$

Change $3\frac{1}{4}$ to an improper fraction and write in lowest terms.

$$3\frac{1}{4} \cdot 12 = \frac{13}{4} \cdot \frac{12}{1} = \frac{13}{\cancel{4}_{1}} \cdot \frac{\cancel{12}^{\,3}}{1} = \frac{39}{1} = 39$$

Show that the cross products are equal.
$$8 \cdot x = 39$$
Divide both sides by 8.

$$\frac{\cancel{8}^{\,1} \cdot x}{\cancel{8}_{\,1}} = \frac{39}{8}$$

Write the solution as a mixed number in lowest terms.

$$x = \frac{39}{8} = 4\frac{7}{8}$$

The unknown number is $4\frac{7}{8}$.

Check in original proportion.

5. Find the unknown number in each proportion.

a. $\dfrac{4\frac{1}{3}}{5} = \dfrac{x}{3}$

_____

$$\frac{3\frac{1}{4}}{8} = \frac{4\frac{7}{8}}{12}$$

$$\nearrow 8 \cdot 4\frac{7}{8} = \frac{\cancel{8}}{1} \cdot \frac{39}{\cancel{8}} = 39$$

$$\searrow 3\frac{1}{4} \cdot 12 = \frac{13}{\cancel{4}} \cdot \frac{\cancel{12}}{1} = 39$$

The cross products are equal, so $4\frac{7}{8}$ is the correct solution.

b. $\dfrac{3.5}{1.4} = \dfrac{4}{x}$

Show that the products are equal.

$$(3.5)(x) = (1.4)(4)$$

$$(3.5)(x) = 5.6$$

Divide both sides by 3.5.

$$\frac{\cancel{(3.5)} \cdot (x)}{\cancel{3.5}} = \frac{5.6}{3.5}$$

$$x = \frac{5.6}{3.5}$$

Complete the division.

$$x = 1.6 \qquad 35\overline{)56.0}^{\,1.6}$$

So, the unknown number is 1.6. Write the solution in the original proportion and check it by finding the cross products.

$$\frac{3.5}{1.4} = \frac{4}{1.6} \quad \nearrow 1.4 \cdot 4 = 5.6$$
$$\searrow 3.5 \cdot 1.6 = 5.6$$

The cross products are equal, so 1.6 is the correct solution.

b. $\dfrac{x}{8} = \dfrac{1.2}{1.5}$

_____

Objective 1   Write proportions.

For extra help, see Example 1 on page 443 of your text and Section Lecture video for Section 6.3 and Exercise Solutions Clip 1, 3, and 5.

Write each proportion.

1.   50 is to 8 as 75 is to 12.

1. _____

2. 36 is to 45 as 8 is to 10.   2. _____

3. 3 is to 33 as 12 is to 132.   3. _____

Objective 2   Determine whether proportions are true or false.

For extra help, see Examples 2–3 on pages 443–445 of your text and Section Lecture video for Section 6.3 and Exercise Solutions Clip 9, 11, 15, and 25.

Determine whether each proportion is true or false by writing the ratios in lowest terms. Show the simplified ratios and then write **true** *or* **false**.

4. $\dfrac{48}{36} = \dfrac{3}{4}$ 4. _____

5. $\dfrac{30}{25} = \dfrac{6}{5}$ 5. _____

Use cross products to determine whether the proportion is true or false. Show the cross products and then write **true** *or* **false**.

6. $\dfrac{4\frac{3}{5}}{9} = \dfrac{18\frac{2}{5}}{36}$ 6. _____

Objective 3   Find the unknown number in a proportion.

For extra help, see Examples 4–5 on pages 446–449 of your text and Section Lecture video for Section 6.3 and Exercise Solutions Clip 31, 33, and 39.

Find the unknown number in the proportion.

7. $\dfrac{9}{7} = \dfrac{x}{28}$ 7. _____

Find the unknown number in each proportion.

8. $\dfrac{2}{3\frac{1}{4}} = \dfrac{8}{x}$ 8. _____

9. $\dfrac{3}{x} = \dfrac{0.8}{5.6}$ 9. _____

Chapter 6 RATIO, PROPORTION, AND LINE/ANGLE/TRIANGLE RELATIONSHIPS

6.4   Problem Solving with Proportions

Learning Objectives
1   Use proportions to solve application problems.

Key Terms

Use the vocabulary terms listed below to complete each statement in exercises 1−2.

rate   ratio

1. A statement that compares a number of inches to a number of inches is a

    _____ .

2. A statement that compares a number of gallons to a number of miles is a

    _____ .

Guided Examples

Review these examples for Objective 1:

1. Alexis drove 343 miles on 7.5 gallons of gas. How far can she travel on a full tank of 12 gallons of gas?

    Step 1 Read the problem. The problem asks for the number of miles the car can travel on 12 gallons of gas.

    Step 2 Work out a plan. Decide what is being compared and write a proportion using the two rates.
    $$\frac{343 \text{ miles}}{7.5 \text{ gallons}} = \frac{x \text{ miles}}{12 \text{ gallons}}$$

    Step 3 Estimate a reasonable answer. To estimate the answer, notice that 12 is a little more than 1.5 times 7.5. So use $343 \cdot 1.5 = 514.5$ miles, as an estimate.

    Step 4 Solve the problem. Ignore the units while solving for x.
    $$\frac{343 \text{ miles}}{7.5 \text{ gallons}} = \frac{x \text{ miles}}{12 \text{ gallons}}$$
    $$(7.5)(x) = (343)(12)$$
    $$(7.5)(x) = 4116$$

Now Try:

1. Aiden spends 23 hours painting 4 apartments. How long will it take him to paint the other 16 apartments?

    _____

$$\frac{\cancel{(7.5)}(x)}{\cancel{7.5}} = \frac{4116}{7.5}$$

$$x = 548.8 \quad \text{Round to } 549.$$

Step 5 State the answer. Rounded to the nearest mile, the car can travel about 549 miles on a full tank of gas.

Step 6 Check your work. The answer 549 miles, is a little more than the estimate of 514.5 miles, so it is reasonable.

2. There are 24 women in a college class of 39. At that rate, how many of the college's 10,400 students are women?

Step 1 Read the problem. The problem asks how many of the 10,400 students are women.

Step 2 Work out a plan. Decide what is being compared and write a proportion using the two rates.

$$\frac{24 \text{ women}}{39 \text{ students}} = \frac{x \text{ women}}{10,400 \text{ students}}$$

Step 3 Estimate a reasonable answer. To estimate the answer, notice that 24 is a little more than half 39. Half of 10,400 is $10,400 \div 2 = 5200$, so our estimate is more than 5200 students.

Step 4 Solve the problem. Ignore the units while solving for *x*.

$$\frac{24 \text{ women}}{39 \text{ students}} = \frac{x \text{ women}}{10,400 \text{ students}}$$
$$39 \cdot x = 24 \cdot 10,400$$
$$39 \cdot x = 249,600$$
$$\frac{\cancel{39} \cdot x}{\cancel{39}} = \frac{249,600}{39}$$
$$x = 6400 \quad \text{No need to round.}$$

Step 5 State the answer. There are 6400 woman at the college.

Step 6 Check your work. The answer 6400 women, is a little more than the estimate of 5200 women, so it is reasonable.

2. A survey showed that 4 out of 5 smokers have tried to quit smoking. At this rate, how many people in a group of 540 have tired to quit smoking?

_____

Objective 1   Use proportions to solve application problems.

For extra help, see Examples 1–2 on pages 456–458 of your text and Section Lecture video for Section 6.4 and Exercise Solutions Clip 5, 17, and 23.

Set up and solve a proportion for each problem.

1. If 22 hats cost $198, find the cost of 12 hats.   1. _____

2. If 150 square yards of carpet cost $3142.50, find the   2. _____
    cost of 210 square yards of the carpet.

3. A biologist tags 50 deer and releases them in a   3. _____
    wildlife preserve area. Over the course of a two-
    week period, she observes 80 deer, of which 12 are
    tagged. What is the estimate for the population of
    deer in this particular area?

Chapter 6 RATIO, PROPORTION, AND LINE/ANGLE/TRIANGLE RELATIONSHIPS

6.5   Geometry: Lines and Angles

Learning Objectives

1	Identify and name lines, line segments, and rays.
2	Identify parallel and intersecting lines.
3	Identify and name angles.
4	Classify angles as right, acute, straight, or obtuse.
5	Identify perpendicular lines.
6	Identify complementary angles and supplementary angles and find the measure of a complement or supplement of a given angle.
7	Identify congruent angles and vertical angles and use this knowledge to find the measures of angles.
8	Identify corresponding angles and alternate interior angles and use this knowledge to find the measures of angles.

Key Terms

Use the vocabulary terms listed below to complete each statement in exercises 1−19.

point	line	line segment	ray
parallel lines	intersecting lines	angle	
degrees	straight angle	right angle	
acute angle	obtuse angle	perpendicular lines	
complementary angles	supplementary angles		
congruent angles	vertical angles		
corresponding angles	alternate interior angles		

1.   A _____ is a part of a line that has one endpoint and which extends infinitely in one direction.

2.   Two lines that intersect to form a right angle are _____.

3.   An angle whose measure is between 90° and 180° is an _____.

4.   A _____ is a location in space.

5.   Two rays with a common endpoint form an _____.

6.   A set of points that form a straight path that extends infinitely in both directions is called a _____.

7.   An angle that measures less than 90° is called an _____.

8.   Angles are measured using _____ .

9.   Two lines in the same plane that never intersect are _____ .

10.   A part of a line with two endpoints is a _____ .

11.   Two lines that cross at one point are _____ .

12.   An angle whose measure is exactly 90° is a _____ .

13.   The nonadjacent angles formed by two intersecting lines are called

         _____ .

14.   Angles whose measures are equal are called _____ .

15.   Two angles whose measures sum to 180° are _____ .

16.   Two angles whose measures sum to 90° are _____ .

17.   When two parallel lines are cut by a transversal, the angles between the parallel lines on opposite sides of the transversal are called

         _____ .

18.   When two parallel lines are cut by a transversal, the angles in the same relative position with regard to the parallel lines and the transversal are called

         _____ .

19.   An angle whose measure is exactly 180° is a _____ .

Guided Examples

Review these examples for Objective 1:

1.   Identify each figure below as a line, line segment, or ray, and name it using the appropriate symbol.

    a.

This figure has two endpoints, so it is a line segment named $\overline{GH}$ or $\overline{HG}$.

Now Try:

1.   Identify each figure below as a line, line segment, or ray, and name it using the appropriate symbol.

    a.

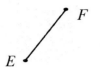

                _____

b.

This figure starts at point B and goes on forever in one direction, so it is a ray named $\overrightarrow{BA}$.

c.

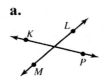

This figure goes on forever in both directions, so it is a line named $\overleftrightarrow{PQ}$ or $\overleftrightarrow{QP}$.

b.

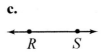

_____

c.

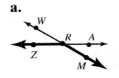

_____

Review these examples for Objective 2:

2. Label each pair of the lines as appearing to be parallel or as intersecting.

    a.

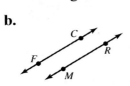

The lines in this figure cross, so they are intersecting lines.

    b.

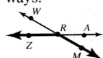

The lines in this figure do not intersect; they appear to be parallel.

Now Try:

2. Label each pair of the lines as appearing to be parallel or as intersecting.

    a.

_____

    b.

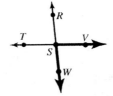

_____

Review this example for Objective 3:

3. Name the highlighted angle in two different ways.

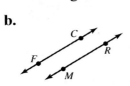

The angle can be named $\angle ZRM$ or $\angle MRZ$. It cannot be named $\angle R$, using the vertex alone, because four different angles have R as their vertex.

Now Try:

3. Name the highlighted angle in two different ways.

_____

Review these examples for Objective 4:

4. Label each angle as acute, right, obtuse, or straight.

a.

This figure shows a straight angle (exactly 180°).

b.

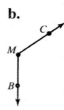

This figure shows an obtuse angle (more than 90° but less than 180°).

c.

This figure shows an acute angle (less than 90°).

d.

This figure shows a right angle (exactly 90° and identified by a small square at the vertex).

Now Try:

4. Label each angle as acute, right, obtuse, or straight.

a.

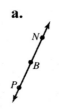

_____

b.

_____

c.

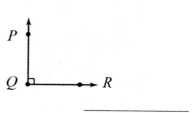

_____

d.

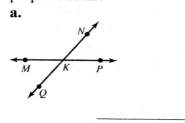

_____

Review these examples for Objective 5:

5. Which pairs of lines are perpendicular?

a.

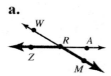

The lines in this figure are intersecting lines, but they are not perpendicular because they do not form a right angle.

Now Try:

5. Which pairs of lines are perpendicular?

a.

_____

b.

b.

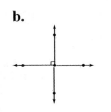

_____

The lines in this figure are perpendicular to each other because they intersect at right angles.

Review these examples for Objective 6:

6. Identify each pair of complementary angles.

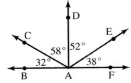

∠*BAC* (32°) and ∠*CAD* (58°) are complementary angles because
$$32° + 58° = 90°$$

∠*DAE* (52°) and ∠*EAF* (38°) are complementary angles because
$$52° + 38° = 90°$$

7. Find the complement of each angle.

a. 66°

Find the complement of 66° by subtracting.
$$90° - 66° = 24°$$

b. 4°

Find the complement of 4° by subtracting.
$$90° - 4° = 86°$$

8. In the figures below, ∠*ABC* measures 150°, ∠*WXY* measures 30°, ∠*MKQ* measures 30°, and ∠*QKP* measures 150°. Identify each pair of supplementary angles.

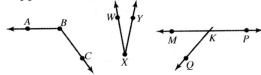

∠*ABC* and ∠*WXY*, because $150° + 30° = 180°$
∠*MKQ* and ∠*QKP*, because $30° + 150° = 180°$
∠*ABC* and ∠*MKQ*, because $150° + 30° = 180°$
∠*WXY* and ∠*QKP*, because $30° + 150° = 180°$

Now Try:

6. Identify each pair of complementary angles.

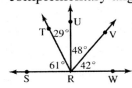

_____

7. Find the complement of each angle.

a. 72°

_____

b. 12°

_____

8. In the figures below, ∠*OQP* measures 35°, ∠*PQN* measures 145°, ∠*RST* measures 35°, and ∠*BMC* measures 145°. Identify each pair of supplementary angles.

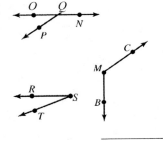

_____

9. Find the supplement of each angle.

    a. 38°

Find the supplement of 38° by subtracting.
    180° − 38° = 142°
    b. 121°

Find the supplement of 121° by subtracting.
    180° − 121° = 59°

9. Find the supplement of each angle.
    a. 82°

    _____

    b. 168°

    _____

Review this example for Objective 7:

10. Identify the angles that are congruent.

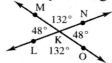

Congruent angles measure the same number of degrees.
$\angle LKM \cong \angle OKN$ and $\angle LKO \cong \angle MKN$.

11. Identify the vertical angles in this figure.

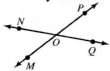

$\angle POQ$ and $\angle NOM$ are vertical angles because they do not share a common side and they are formed by two intersecting lines $\left(\overleftrightarrow{NQ} \text{ and } \overleftrightarrow{MP} \right)$.

$\angle NOP$ and $\angle MOQ$ are also vertical angles.

Now Try:

10. Identify the angles that are congruent.

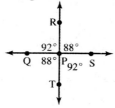

    _____

11. Identify the vertical angles in this figure.

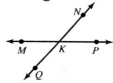

    _____

12. In the figure below, $\angle APB$ measures 93° and $\angle BPC$ measures 37°. Find the measures of the indicated angles.

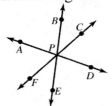

a. $\angle EPF$

$\angle EPF$ and $\angle BPC$ are vertical angles, so they are congruent. This means they measure the same number of degrees.
The measure of $\angle BPC$ is 37°, so the measure of $\angle EPF$ is 37° also.

b. $\angle DPE$

$\angle DPE$ and $\angle APB$ are vertical angles, so they are congruent.
The measure of $\angle APB$ is 93°, so the measure of $\angle DPE$ is 93° also.

c. $\angle CPD$

Look at $\angle CPD$, $\angle BPC$, and $\angle APB$. Notice that $\overrightarrow{PA}$ and $\overrightarrow{PD}$ go in opposite directions. Therefore, $\angle APD$ is a straight angle and measures 180°. To find the measure of $\angle CPD$, subtract the sum of the other two angles from 180°.

$$180° - (93° + 37°) = 180° - (130°) = 50°$$

The measure of $\angle CPD$ is 50°.

d. $\angle FPA$

$\angle FPA$ and $\angle CPD$ are vertical angles, so they are congruent. We know from part (c) that the measure of $\angle CPD$ is 50°, so the measure of $\angle FPA$ is 50° also.

12. In the figure below, $\angle AGF$ measures 33° and $\angle BGC$ measures 105°. Find the measures of the indicated angles.

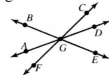

a. $\angle CGD$

_____

b. $\angle EGF$

_____

c. $\angle DGE$

_____

d. $\angle BGA$

_____

Name: Date:
Instructor: Section:

Review these examples for Objective 8:

13. In each figure, line *m* is parallel to line *n*.

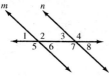

a. Identify all pairs of corresponding angles.

There are four pair of corresponding angles:

∠1 and ∠3   ∠2 and ∠4
∠5 and ∠7   ∠6 and ∠8

b. Identify all pairs of alternate interior angles.

There are two pair of alternate interior angles:
∠2 and ∠7
∠6 and ∠3

14. In the figure below, line *m* is parallel to line *n* and the measure of ∠3 is 44°. Find the measure of the other angles.

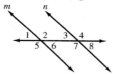

We know that ∠3 is 44°.
   ∠3 ≅ ∠1 (corresponding angles), so the measure of ∠1 is also 44°.
   ∠3 ≅ ∠6 (alternate interior angles), so the measure of ∠6 is also 44°.
   ∠6 ≅ ∠8 (corresponding angles), so the measure of ∠8 is also 44°.
Notice that the exterior sides of ∠3 and ∠4 form a straight line. Therefore, ∠3 and ∠4 are supplementary angles. If ∠3 is 44°, then ∠4 must be $180° - 44° = 136°$.
   ∠4 ≅ ∠2 (corresponding angles), so the measure of ∠4 is also 136°.
   ∠7 ≅ ∠2 (alternate interior angles), so the measure of ∠2 is also 136°.
   ∠7 ≅ ∠5 (corresponding angles), so the measure of ∠5 is also 136°.

Now Try:

13. In each figure, line *m* is parallel to line *n*. Identify all pairs of corresponding angles and all pairs of alternate interior angles.

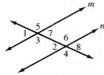

a. Identify all pairs of corresponding angles.

_____

_____

b. Identify all pairs of alternate interior angles.

_____

_____

14. In the figure below, line *m* is parallel to line *n* and the measure of ∠6 is 125°. Find the measure of the other angles.

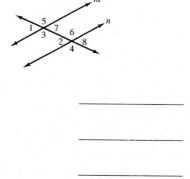

_____

_____

_____

_____

Copyright © 2014 Pearson Education, Inc.

Name: Date:

Instructor: Section:

Objective 1   Identify and name lines, line segments, and rays.

For extra help, see Example 1 on page 463 of your text and Section Lecture video for Section 6.5.

Identify each figure as a line, line segment, or ray, and name it.

1.

2.

3.

1. _____

2. _____

3. _____

Objective 2   Identify parallel and intersecting lines.

For extra help, see Example 2 on page 464 of your text and Section Lecture video for Section 6.5.

Label each pair of lines as appearing to be **parallel** *or* **intersecting**.

4.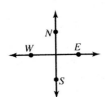

5.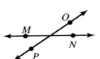

6.

4. _____

5. _____

6. _____

219

Name: Date:

Instructor: Section:

Objective 3   Identify and name angles.

For extra help, see Example 3 on page 465 of your text and Section Lecture video for Section 6.5 and Exercise Solutions Clip 15.

Name each angle drawn with darker rays by using the three-letter form of identification.

7.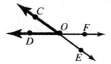

                                                        7. _____

8.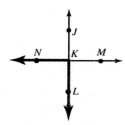

                                                        8. _____

9.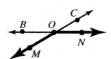

                                                        9. _____

Objective 4   Classify angles as right, acute, straight, or obtuse.

For extra help, see Example 4 on page 466 of your text and Section Lecture video for Section 6.5.

Label each angle as **acute**, **right**, **obtuse**, *or* **straight**.

10.

                                                        10. _____

11.

                                                        11. _____

12.

                                                        12. _____

Name: Date:
Instructor: Section:

Objective 5   Identify perpendicular lines.

For extra help, see Example 5 on page 467 of your text and Section Lecture video for Section 6.5.

Label each pair of lines as appearing to be **parallel**, **perpendicular**, *or* **intersecting**.

13.

13. _____

14.

14. _____

15.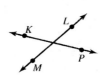

15. _____

Objective 6   Identify complementary angles and supplementary angles and find the *measure* of a complement or supplement of a given angle.

For extra help, see Examples 6–9 on pages 467–468 of your text and Section Lecture video for Section 6.5 and Exercise Solutions Clip 25 and 27.

Find the complement of the angle.

16.   43°

16. _____

Find the supplement of the angle.

17.   16°

17. _____

Identify each pair of supplementary angles.

18.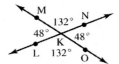

18. _____

**Objective 7   Identify congruent angles and vertical angles and use this knowledge
to find the measures of angles.**

For extra help, see Examples 10–12 on pages 469–470 of your text and Section Lecture
video for Section 6.5 and Exercise Solutions Clip 37.

In the figure below, identify the angles that are congruent.

19.   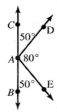   19. _____

In the figure below, identify all the vertical angles.

20.   20. _____

*In the figure below, ∠ABE measures 73° and ∠FEB measures 107°. Find the measures
of the indicated angles.*

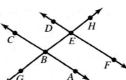

21.   ∠CBG and ∠DEH   21. _____

Name: Date:
Instructor: Section:

**Objective 8   Identify corresponding angles and alternate interior angles and use
this knowledge to find the measures of angles.**

For extra help, see Examples 13–14 on pages 471–472 of your text and Section Lecture
video for Section 6.5 and Exercise Solutions Clip 47 and 53.

*In each figure, line m is parallel to line n. List the corresponding angles and the alternate
interior angles. Then find the measure of each angle.*

22.   ∠4 measures 100°.

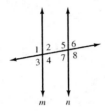

22.
Corresponding angles:

_____

Alternate interior angles:

_____

Angle measures:

_____

23.   ∠7 measures 37°.

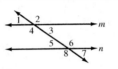

23.
Corresponding angles:

_____

Alternate interior angles:

_____

Angle measures:

_____

223

Chapter 6 RATIO, PROPORTION, AND LINE/ANGLE/TRIANGLE RELATIONSHIPS

6.6   Geometry Applications: Congruent and Similar Triangles

Learning Objectives	
1	Identify corresponding parts of congruent triangles.
2	Prove that triangles are congruent using SAS, SSS, or ASA.
3	Identify corresponding parts of similar triangles.
4	Find the unknown lengths of sides in similar triangles.
5	Solve application problems involving similar triangles.

Key Terms

Use the vocabulary terms listed below to complete each statement in exercises 1–4.

congruent figures   **similar figures**

congruent triangles   **similar triangles**

1.   Triangles with the same shape and size are _____.

2.   _____ are identical both in shape and in size.

3.   Triangles with the same shape but not necessarily the same size are

_____.

4.   _____ have the same shape but are different sizes.

Guided Examples

Review this example for Objective 1:

1.   Identify corresponding angles and sides in these congruent triangles.

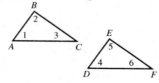

The corresponding parts are:

∠1 and ∠4 $\overline{AB}$ and $\overline{DE}$

∠2 and ∠5 $\overline{BC}$ and $\overline{EF}$

∠3 and ∠6 $\overline{AC}$ and $\overline{DF}$

Now Try:

1.   Identify corresponding angles and sides in these congruent triangles.

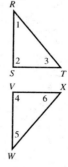

_____

Name: Date:

Instructor: Section:

Review these examples for Objective 2:

2. Explain which method can be used to prove that each pair of triangles is congruent. Choose from ASA, SSS, and SAS.

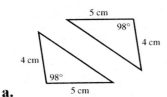

a.

On both triangles, two corresponding sides and the angle between them measure the same, so Side-Angle-Side (SAS) method can be used to prove that the triangles are congruent.

b.

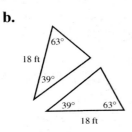

On both triangles, two corresponding angles and the side between them measure the same, so the Angle-Side-Angle (ASA) method can be used to prove that the triangles are congruent.

c.

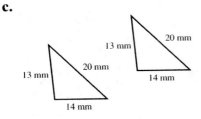

Each pair of corresponding sides has the same length, so the Side-Side-Side (SSS) method can be used to prove that the triangles are congruent.

Now Try:

2. Explain which method can be used to prove that each pair of triangles is congruent. Choose from ASA, SSS, and SAS.

a.

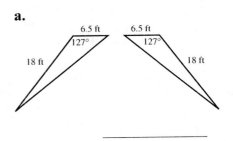

_____

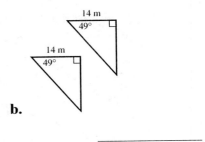

b.

_____

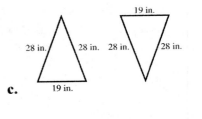

c.

_____

Name: Date:

Instructor: Section:

Review this example for Objective 3:

3. Identify corresponding angles and sides in these similar triangles.

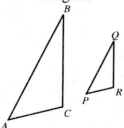

$\overline{AB}$ and $\overline{PQ}$; $\overline{AC}$ and $\overline{PR}$; $\overline{BC}$ and $\overline{QR}$;

$\angle A$ and $\angle P$; $\angle B$ and $\angle Q$; $\angle C$ and $\angle R$

Now Try:

3. Identify corresponding angles and sides in these similar triangles.

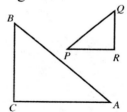

_____

Review these examples for Objective 4:

4. Find the length of b in the smaller triangle. Assume the triangles are similar.

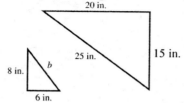

The length you want to find in the smaller triangle is side b, and it corresponds to 25 in. in the larger triangle. The smaller triangle is turned "upside down" compared to the larger triangle, so be careful when identifying corresponding sides. Then notice that 6 in. in the smaller triangle corresponds to 15 in. in the larger triangle, and you know both of their lengths. Because the ratios of the lengths of corresponding sides are equal, you can set up a proportion.

$$\frac{b}{25} = \frac{6}{15}$$

$$\frac{b}{25} = \frac{2}{5} \quad \text{Write } \frac{6}{15} \text{ in lowest terms as } \frac{2}{5}.$$

Find the cross products.

$$b \cdot 5 = 25 \cdot 2$$

$$\frac{b \cdot \cancel{5}^{1}}{\cancel{5}_{1}} = \frac{50}{5}$$

$$b = 10$$

Side b has length of 10 in.

Now Try:

4. Find the length of x in the smaller triangle. Assume the triangles are similar.

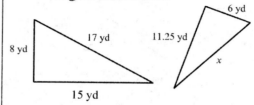

_____

5. Find the perimeter of the smaller triangle. Assume the triangles are similar.

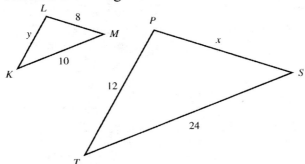

5. Find the perimeter of the smaller triangle. Assume the triangles are similar.

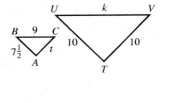

_____

First, find y, the length of $\overline{KL}$, then add the lengths of all three sides to find the perimeter. The ratios of the lengths of corresponding sides are equal, so you can set up a proportion.

$$\begin{array}{c} KL \to \\ KM \to \end{array} \frac{y}{10} = \frac{12}{24} \begin{array}{c} \leftarrow PT \\ \leftarrow TS \end{array}$$

$$\frac{y}{10} = \frac{1}{2} \quad \text{Write } \frac{12}{24} \text{ in lowest terms as } \frac{1}{2}.$$

Find the cross products.

$$\frac{y}{10} = \frac{1}{2}$$

$$y \cdot 2 = 10 \cdot 1$$

$$\frac{y \cdot \overset{1}{\cancel{2}}}{\cancel{2}_{1}} = \frac{10}{2}$$

$$y = 5$$

$\overline{KL}$ has a length of 5.

Now add the lengths of all three sides to find the perimeter of the smaller triangle.

$$\text{Perimeter} = 8 + 10 + 5 = 23$$

Review this example for Objective 5:

6. The height of the house shown here can be found by using similar triangles and proportion. Find the height of the house by writing a proportion and solving it.

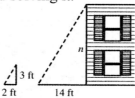

The triangles shown are similar, so write a proportion to find *n*.

height of larger triangle → $\dfrac{n}{3}$ = $\dfrac{14}{2}$ ← length of larger triangle
height of smaller triangle → $\quad\quad$ ← length of smaller triangle

Find the cross products and show that they are equal.

$$n \cdot 2 = 3 \cdot 14$$
$$n \cdot 2 = 42$$
$$\frac{n \cdot \cancel{2}}{\cancel{2}} = \frac{42}{2}$$
$$n = 21$$

The height of the house is 21 ft.

Now Try:

6. A flagpole casts a shadow 77 feet long at the same time that a pole 15 feet tall casts a shadow 55 ft long. Find the height of the flagpole.

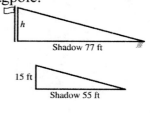

_____

Objective 1   Identify corresponding parts of congruent triangles.

For extra help, see Example 1 on pages 477–478 of your text and Section Lecture video for Section 6.6.

Each pair of triangles is congruent. List the corresponding angles and the corresponding sides.

1.

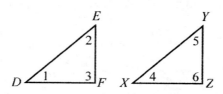

1. _____

2.

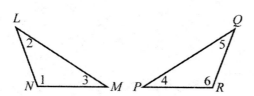

2. _____

Objective 2   Prove that triangles are congruent using SAS, SSS, and ASA.

For extra help, see Example 2 on page 479 of your text and Section Lecture video for Section 6.6.

Determine which of these methods can be used to prove that each pair of triangles is congruent: Angle-Side-Angle (ASA), Side-Side-Side (SSS), or Side-Angle-Side (SAS).

3.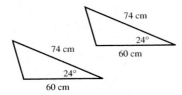

3. _____

4.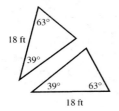

4. _____

Objective 3   Identify the corresponding parts of similar triangles.

For extra help, see page 479 of your text and Section Lecture video for Section 6.6.

Name the corresponding angles and the corresponding sides in each pair of similar triangles.

5.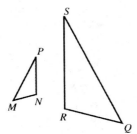

5. _____

6.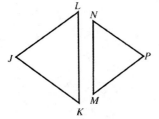

6. _____

7. 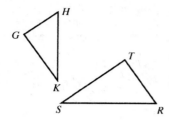   7. _____

Objective 4   Find the unknown lengths of sides in similar triangles.

For extra help, see Examples 3–4 on pages 480–481 of your text and Section Lecture video for Section 6.6 and Exercise Solutions Clip 21 and 23.

Find the unknown lengths in each pair of similar triangles.

8. 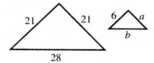   8. _____

9. 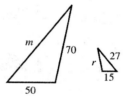   9. _____

Find the perimeter of each triangle. Assume the triangles are similar.

10. 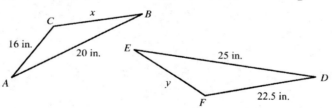   **10.** *ABC* _____

                                                                            EDF _____

Name: Date:

Instructor: Section:

Objective 5   Solve application problems involving similar triangles.

For extra help, see Example 5 on page 482 of your text and Section Lecture video for Section 6.6 and Exercise Solutions Clip 29.

Solve each application problem.

11.   A sailor on the USS Ramapo saw one of the highest waves ever recorded. He used the height of the ship's mast, the length of the deck and similar triangles to find the height of the wave. Using the information in the figure, write a proportion and then find the height of the wave.

     11.

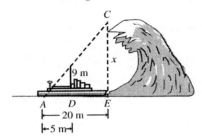

12.   A fire lookout tower provides an excellent view of the surrounding countryside. The height of the tower can be found by lining up the top of the tower with the top of a 3-meter stick. Use similar triangles to find the height of the tower.

     12. _____

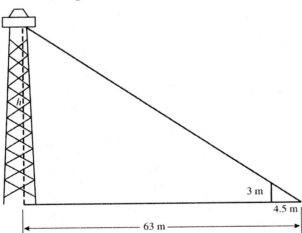

13. A 30 m ladder touches the side of a building at a   **13.** _____
height of 25 m. At what height would a 12-m ladder
touch the building if it makes the same angle with
the ground?

Chapter 7 PERCENT

7.1   The Basics of Percent

Learning Objectives
1   Learn the meaning of percent.
2   Write percents as decimals.
3   Write decimals as percents.
4   Write percents as fractions.
5   Write fractions as percents.
6   Use 100% and 50%.

Key Terms

Use the vocabulary terms listed below to complete each statement in exercises 1–3.

percent   **ratio**   **decimals**

1. To compare two quantities that have the same type of units, use a _____.

2. _____ means per one hundred.

3. _____ represent parts of a whole.

Guided Examples

Review these examples for Objective 1:

1. Write a percent to describe each situation.

   a. You leave a $25 tip when the restaurant bill was $100. What percent is the tip?

   The tip is $25 per $100, or $\frac{25}{100}$. The percent is 25%.

   b. You pay $8 tax on a $100 catering order. What is the tax rate?

   The tax is $8 per $100, or $\frac{8}{100}$. The percent is 8%.

   c. You earn 91 points on a 100-point math test. What percent is your score?

   Then your score is 91 per 100, or $\frac{91}{100}$, or 91%.

Now Try:

1. Write a percent to describe each situation.

   a. You give $10 to a charity per $100. What is the percent given to charity?

   _____

   b. The tax on a hotel room of $100 is $18. What is the tax rate?

   _____

   c. You earn 85 points on a 100-point math quiz. What percent is your score?

   _____

Name: Date:

Instructor: Section:

Review these examples for Objective 2:

2. Write each percent as a decimal.

   a. 39%

   $39\% = 39 \div 100 = 0.39$

   b. 7%

   $7\% = 7 \div 100 = 0.07$

   c. 89.9%

   $89.9\% = 89.9 \div 100 = 0.899$

   d. 400%

   $400\% = 400 \div 100 = 4.00$

   e. 169%

   $169\% = 169 \div 100 = 1.69$

3. Write each percent as a decimal by dropping the percent symbol and moving the decimal point two places to the left.

   a. 56%

   Drop the percent sign and move the decimal point two places to the left.
      $56\% = 56.\% = 0.56$

   b. 120%

   $120\% = 120.\% = 1.20$ or 1.2

   c. 1.3%

   0 is attached so the decimal point can be moved two places to the left.
      $1.3\% = 0.013$

   d. 0.2%

   Two zeros are attached so the decimal point can be moved two places to the left.
      $0.2\% = 0.002$

Now Try:

2. Write each percent as a decimal.

   a. 23%

   _____

   b. 8%

   _____

   c. 15.3%

   _____

   d. 500%

   _____

   e. 302%

   _____

3. Write each percent as a decimal by dropping the percent symbol and moving the decimal point two places to the left.

   a. 48%

   _____

   b. 240%

   _____

   c. 2.5%

   _____

   d. 0.8%

   _____

Review these examples for Objective 3:

4. Write each decimal as a percent by moving the decimal point two places to the right.

   a. 0.31

   Decimal point is moved two places to the right and percent symbol is attached.
   $0.31 = 31\%$

   b. 1.6

   0 is attached so the decimal point can be moved two places to the right.
   $1.6 = 1.60 = 160\%$

   c. 0.904

   $0.904 = 90.4\%$

Now Try:

4. Write each decimal as a percent by moving the decimal point two places to the right.

   a. 0.43

   _____

   b. 2.3

   _____

   c. 0.751

   _____

Review these examples for Objective 4:

5. Write each percent as a fraction or mixed number in lowest terms.

   a. 12%

   Recall writing 12% as a decimal.
   $12\% = 12 \div 100 = 0.12$
   Because 0.12 means 12 hundredths,
   $$0.12 = \frac{12}{100} = \frac{12 \div 4}{100 \div 4} = \frac{3}{25}$$

   b. 86%

   Write 86% as $\frac{86}{100}$.

   Write $\frac{86}{100}$ in lowest terms.
   $$\frac{86}{100} = \frac{86 \div 2}{100 \div 2} = \frac{43}{50}$$

   c. 350%

   $$350\% = \frac{350}{100} = \frac{350 \div 50}{100 \div 50} = \frac{7}{2} = 3\frac{1}{2}$$

Now Try:

5. Write each percent as a fraction or mixed number in lowest terms.

   a. 30%

   _____

   b. 78%

   _____

   c. 175%

   _____

6. Write each percent as a fraction in lowest terms.

a. 62.5%

Write 62.5 over 100.

$$62.5\% = \frac{62.5}{100}$$

To get a whole number in the numerator, multiply the numerator and denominator by 10.

$$\frac{62.5}{100} = \frac{62.5(10)}{100(10)} = \frac{625}{1000}$$

Now write the fraction in lowest terms.

$$\frac{625}{1000} = \frac{625 \div 125}{1000 \div 125} = \frac{5}{8}$$

b. $16\frac{2}{3}\%$

Write $16\frac{2}{3}\%$ over 100.

$$16\frac{2}{3}\% = \frac{16\frac{2}{3}}{100}$$

When there is a mixed number in the numerator, rewrite the mixed number as an improper fraction.

$$\frac{16\frac{2}{3}}{100} = \frac{\frac{50}{3}}{100}$$

Next, rewrite the division problem in a horizontal form. Finally, multiply by the reciprocal of the divisor.

$$\frac{\frac{50}{3}}{100} = \frac{50}{3} \div 100 = \frac{50}{3} \div \frac{100}{1} = \frac{50}{3} \cdot \frac{1}{100} = \frac{1}{6}$$

6. Write each percent as a fraction in lowest terms.

a. 43.6%

_____

b. $22\frac{2}{9}\%$

_____

Review these examples for Objective 5:

7. Write each fraction as a percent. Round to the nearest tenth as necessary.

a. $\frac{3}{5}$

$$\frac{3}{5} = \left(\frac{3}{5}\right)(100\%) = \left(\frac{3}{5}\right)\left(\frac{100}{1}\%\right) = \left(\frac{3}{5}\right)\left(\frac{20 \cdot 5}{1}\%\right)$$

$$= \frac{60}{1}\% = 60\%$$

Now Try:

7. Write each fraction as a percent. Round to the nearest tenth as necessary.

a. $\frac{3}{20}$

_____

b. $\dfrac{5}{16}$

$$\dfrac{5}{16} = \left(\dfrac{5}{16}\right)(100\%) = \left(\dfrac{5}{16}\right)\left(\dfrac{100}{1}\%\right)$$

$$= \left(\dfrac{5}{4\cdot 4}\right)\left(\dfrac{4\cdot 25}{1}\%\right)$$

$$= \dfrac{125}{4}\% = 31\dfrac{1}{4}\%$$

c. $\dfrac{1}{15}$

$$\dfrac{1}{15} = \left(\dfrac{1}{15}\right)(100\%) = \left(\dfrac{1}{15}\right)\left(\dfrac{100}{1}\%\right)$$

$$= \left(\dfrac{1}{3\cdot 5}\right)\left(\dfrac{5\cdot 20}{1}\%\right)$$

$$= \dfrac{20}{3}\% = 6\dfrac{2}{3}\%$$

$6\dfrac{2}{3}\%$ rounds to 6.7%.

Review these examples for Objective 6:

8. Fill in the blanks.

   a. 100% of $69 is _____ .

   100% is all of the money. So 100% of $69 is $69.

   b. 100% of 25 miles is _____ .

   100% is all of the miles. So 100% of 25 miles is 25 miles.

9. Fill in the blanks.

   a. 50% of $68 is _____ .

   50% is half of the money. So 50% of $68 is $34.

   b. 50% of 13 hours is _____ .

   50% is half of the hours. So 50% of 13 hours is $6\dfrac{1}{2}$ hours.

b. $\dfrac{15}{16}$

_____

c. $\dfrac{1}{18}$

_____

Now Try:

8. Fill in the blanks.

   a. 100% of $6.15 is _____ .

   _____

   b. 100% of $23\dfrac{1}{2}$ days is _____ .

   _____

9. Fill in the blanks.

   a. 50% of $7.30 is _____ .

   _____

   b. 50% of 19 pages is _____ .

   _____

Objective 1   Learn the meaning of percent.

For extra help, see Example 1 on page 506 of your text and Section Lecture video for Section 7.1

Write as a percent.

1.   68 people out of 100 drive small cars.   1. _____

2.   The tax is $8 per $100.   2. _____

3.   The cost for labor was $45 for every $100 spent to manufacture an item.   3. _____

Objective 2   Write percents as decimals.

For extra help, see Examples 2–3 on pages 507–508 of your text and Section Lecture video for Section 7.1 and Exercise Solutions Clip 5, 11, and 17.

Write each percent as a decimal.

4.   42%   4. _____

5.   310%   5. _____

6.   18.9%   6. _____

Objective 3   Write decimals as percents.

For extra help, see Example 4 on pages 508–509 of your text and Section Lecture video for Section 7.1 and Exercise Solutions Clip 19, 21, 27, and 31.

Write each decimal as a percent.

7.   0.2   7. _____

8.   0.564   8. _____

9.   4.93   9. _____

Objective 4   Write percents as fractions.

For extra help, see Examples 5–6 on pages 509–511 of your text and Section Lecture video for Section 7.1 and Exercise Solutions Clip 39 and 47.

Write each percent as a fraction or mixed number in lowest terms.

10.   140%   10. _____

11. $18\frac{1}{3}\%$

11. _____

12. 55.6%

12. _____

Objective 5 Write fractions as percents.

For extra help, see Example 7 on pages 511–513 of your text and Section Lecture video for Section 7.1.

Write each fraction or mixed number as a percent. If you're using a calculator, first work each one by hand. Then use your calculator and round to the nearest tenth of a percent, if necessary.

13. $\dfrac{47}{50}$

13. _____

14. $\dfrac{11}{40}$

14. _____

15. $\dfrac{64}{75}$

15. _____

Objective 6 Use 100% and 50%.

For extra help, see Examples 8–9 on page 513 of your text and Section Lecture video for Section 7.1 and Exercise Solutions Clip 103.

Fill in the blanks. Remember that 100% is all of something and 50% is half of it.

16. 50% of 260 miles is _____ .

16. _____

17. 100% of $520 is _____ .

17. _____

18. 50% of 40 acres is _____ .

18. _____

Chapter 7 PERCENT

7.2   The Percent Proportion

Learning Objectives
1   Identify the percent, whole, and part.
2   Solve percent problems using the percent proportion.

Key Terms

Use the vocabulary terms listed below to complete each statement in exercises 1–3.

percent proportion   whole   part

1.   The _____ in a percent problem is the entire quantity.

2.   The _____ in a percent problem is the portion being compared with the whole.

3.   Part is to whole as percent is to 100 is called the _____.

Guided Examples

Review these examples for Objective 1:
1.   Find the percent in the following.

   a. 43% of 1100 people exercise daily.

   The percent is 43.

   b. $255 is 29 percent of what number?

   The percent is 29, because 29 appears with the word percent.

   c. What percent of 6500 tons is 275 tons?

   The word percent has no number with it, so the percent is the unknown part of the problem.

Now Try:
1.   Find the percent in the following.
   a. 950 tons of trash is 59% of what number of tons?

   _____

   b. Find the amount of sales tax by multiplying $650 and 7.75 percent.

   _____

   c. If 2650 children attend preschool, what percent of 3280 children attend preschool?

   _____

2. Identify the whole in the following.

    a. 43% of 1100 people exercise daily.

    The whole is 1100.

    b. $255 is 29 percent of what number?

    The whole is the unknown part of the problem.

    c. What percent of 6500 tons is 275 tons?

    The whole is 6500.

3. Identify the part in the following.

    a. 43% of 1100 people exercise daily.

    The part is the unknown part of the problem.

    b. $255 is 29 percent of what number?

    The part is $255.

    c. What percent of 6500 tons is 275 tons?

    The part is 275.

2. Identify the whole in the following.

    a. Find the amount of sales tax by multiplying $650 and 7.75 percent.

    _____

    b. 950 tons of trash is 59% of what number of tons?

    _____

    c. If 2650 children attend preschool, what percent of 3280 children attend preschool?

    _____

3. Identify the part in the following.

    a. Find the amount of sales tax by multiplying $650 and 7.75 percent.

    _____

    b. 950 tons of trash is 59% of what number of tons?

    _____

    c. If 2650 children attend preschool, what percent of 3280 children attend preschool?

    _____

Review these examples for Objective 2:

4. Find 24% of $650.

Here the percent is 24 and the whole is $650. Now find the part. Let x represent the unknown part.

$$\frac{x}{650} = \frac{24}{100} \quad \text{or} \quad \frac{x}{650} = \frac{6}{25}$$

Find the cross products in the proportion and show that they are equal.

Now Try:

4. Find 38% of 6500.

    _____

$$x \cdot 25 = 650 \cdot 6$$

$$x \cdot 25 = 3900$$

$$\frac{x \cdot 25}{25} = \frac{3900}{25}$$

$$x = 156$$

24% of $650 is $156.

5. Use the percent proportion to answer this question.

   9 pounds is what percent of 450 pounds?

The percent proportion is $\dfrac{\text{percent}}{100} = \dfrac{\text{part}}{\text{whole}}$. Set up the proportion using p as the variable representing the unknown percent. Then find the cross products.

$$\frac{p}{100} = \frac{9}{450}$$

$$p \cdot 450 = 900$$

$$\frac{p \cdot \cancel{450}}{\cancel{450}} = \frac{900}{450}$$

$$p = 2$$

The percent is 2%. So 9 pounds is 2% of 450 pounds.

6. Use the percent proportion to answer this question.

   150 students is 75% of how many students?

The percent proportion is $\dfrac{\text{percent}}{100} = \dfrac{\text{part}}{\text{whole}}$.

Here the percent is 75%, the part is 150, and the whole is unknown.

$$\frac{75}{100} = \frac{150}{n}$$

$$75 \cdot n = 15,000$$

$$\frac{\cancel{75} \cdot n}{\cancel{75}} = \frac{15,000}{75}$$

$$n = 200$$

The whole is 200 students. So 150 students is 75% of 200 students.

5. Use the percent proportion to answer this question. $28 is what percent of $700?

   _____

6. Use the percent proportion to answer this question. 45 cars is 30% of how many cars?

   _____

7. Use the percent proportion to answer each question.

a. How many cats is 275% of 20 cats?

Here, the percent is 275%, the part is unknown, and the whole is 20. Use the percent proportion.

$$\frac{275}{100} = \frac{n}{20}$$

$$100 \cdot n = 5500$$

$$\frac{\cancel{100} \cdot n}{\cancel{100}} = \frac{5500}{100}$$

$$n = 55$$

The part is 55 cats.

b. What percent of $40 is $46?

Here, the percent is unknown, the part is $46, and the whole is $40. Use the percent proportion.

$$\frac{p}{100} = \frac{46}{40}$$

$$p \cdot 40 = 4600$$

$$\frac{p \cdot \cancel{40}}{\cancel{40}} = \frac{4600}{40}$$

$$p = 115$$

The percent is 115%.

7. Use the percent proportion to answer each question.

a. 66 days is what percent of 30 days?

_____

b. 185% of what amount is $173?

_____

Objective 1   Identify the percent, whole, and part.

For extra help, see Examples 1–3 on pages 521–522 of your text and Section Lecture video for Section 7.2 and Exercise Solutions Clip 5.

In each of the following, identify the percent, the whole, and the part. Do not try to solve for any unknowns.

1. $50 is 250% of what number?

1. percent _____

   whole _____

   part _____

2. 336 students is what percent of 840 students?

2. percent _____

   whole _____

   part _____

3. The state sales tax is $7\frac{3}{4}$ percent of the $895 price.

    3. percent _____

          whole _____

          part _____

Objective 2   Solve percent problems using the percent proportion.

For extra help, see Examples 4–7 on pages 522–525 of your text and Section Lecture video for Section 7.2 and Exercise Solutions Clip 7, 9, and 21.

Use the percent proportion to answer these questions. If necessary, round money answers to the nearest cent and percent answers to the nearest tenth of a percent.

4. 165% of what number of feet is 12.7 feet? Round the   **4.** _____
      answer to the nearest tenth.

5. What percent of 78 bicyclists is 110 bicyclists?   **5.** _____

6. What number is 43% of 2200?   **6.** _____

Chapter 7 PERCENT

7.3   The Percent Equation

Learning Objectives
1   Estimate answers to percent problems involving 25%.
2   Find 10% and 1% of a number by moving the decimal point.
3   Solve basic percent problems using the percent equation.

Key Terms

Use the vocabulary terms listed below to complete each statement in exercises 1–2.

    percent equation   **percent**

1.   A number written with a _____ sign means "divided by 100".

2.   The _____ is part = percent · whole .

Guided Examples

Review these examples for Objective 1:
1.   Estimate the answer to each question.

   a. What is 25% of $423?

   Use front end rounding to round $423 to $400.
   Then divide $400 by 4. The estimate is $100.

   b. Find 25% of 82.7 miles.

   Use front end rounding to round 82.7 miles
   to 80 miles. Then divide 80 by 4. The
   estimate is 20 miles.

   c. 25% of 89 days is how long?

   Round 89 days to 90 days using front end
   rounding. Then divide 90 by 4 to get
   22.5 days.

Now Try:
1.   Estimate the answer to each
   question.
   a. Find 25% of $205.

   _____

   b. What is 25% of 59 hours?

   _____

   c. 25% of 97 pounds is how
   many pounds?

   _____

Review these examples for Objective 2:
2.   Find the exact answer to each question by
   moving the decimal point.

   a. What is 10% of $674?

   To find 10% of $674, divide $674 by 10. Do the
   division by moving the decimal point one place
   to the left.
      10% of $674 = $67.40.

Now Try:
2.   Find the exact answer to each
   question by moving the decimal
   point.
   a. What is 10% of $932?

   _____

b. Find 10% of 37.6 miles.

To find 10% of 37.6 miles, divide 37.6 by 10. Move the decimal point one place to the left.
10% of 37.6 miles = 3.76 miles

3. Find the exact answer to each question by moving the decimal point.

a. What is 1% of $947?

To find 1% of $947, divide $947 by 100. Do the division by moving the decimal point two places to the left.
1% of $947 = $9.47

b. Find 1% of 68.9 miles.

To find 1% of 68.9 miles, divide 68.9 by 100. Move the decimal point two places to the left.
1% of 68.9 = 0.689 mile

b. Find 10% of 693 miles.

_____

3. Find the exact answer to each question by moving the decimal point.
a. What is 1% of $549?

_____

b. Find 1% of 30.7 miles.

_____

Review these examples for Objective 3:
4. Write and solve a percent equation to answer each question.

a. 35% of $245 is how much money?

Translate the sentence into an equation, where *of* indicates multiplication and *is* translates to the equal sign. The percent must be written in decimal form.

35% of $245 is how much money?

$$0.35 \cdot 245 = n$$

To solve, simplify the left side.
$$(0.35)(245) = n$$
$$85.75 = n$$
So, 35% of $245 is $85.75.

b. How many books is 310% of 70 books?

Translate the sentence into an equation. Write the percent in decimal form.
How many books is 310% of 70 books?

$$n = 3.10 \cdot 70$$

This time the two sides of the percent equation are reversed, so

Now Try:
4. Write and solve a percent equation to answer each question.
a. 89% of $310 is how much money?

_____

b. How many miles is 180% of 60 miles?

_____

part = percent · whole

To solve the equation, simplify the right side, multiplying 3.10 by 70.

$$n = (3.10)(70)$$

$$n = 217$$

So 217 books is 310% of 70 books.

5. Write and solve a percent equation to answer each question.

a. 16 pounds is what percent of 200 pounds?

Translate the sentence into an equation. This time the percent is unknown.

16 pounds   is   what percent   of   200 pounds?

$$\begin{array}{ccccc} \downarrow & \downarrow & \downarrow & \downarrow & \downarrow \\ 16 & = & p & \cdot & 200 \end{array}$$

To solve the equation, divide both sides by 200.

$$\frac{16}{200} = \frac{p \cdot \cancel{200}}{\cancel{200}}$$

$$0.08 = p$$

$$0.08 = 8\%$$

So 16 pounds is 8% of 200 pounds.

b. What percent of $80 is $132?

Translate the sentence into an equation and solve it.

What percent   of $80   is $132

$$\begin{array}{ccccc} \downarrow & & \downarrow & \downarrow & \downarrow \\ p & \cdot & 80 & = & 132 \end{array}$$

To solve, divide both sides by 80.

$$\frac{p \cdot \cancel{80}}{\cancel{80}} = \frac{132}{80}$$

$$p = 1.65$$

$$1.65 = 165\%$$

So 165% of $80 is $132.

5. Write and solve a percent equation to answer each question.

a. 66 hours is what percent of 30 hours?

_____

b. What percent of 100 is 33?

_____

6. Write and solve a percent equation to answer each question.

  a. 126 credits is 70% of how many credits?

Translate the sentence into an equation. Write the percent in decimal form.

$\underbrace{126 \text{ credits}}$ is   70%   of $\underbrace{\text{how many credits?}}$

$$126 \quad = \quad 0.70 \quad \cdot \quad n$$

Recall that 0.70 is equivalent to 0.7, so use 0.7 in the equation. To solve, divide both sides by 0.7.

$$\frac{126}{0.7} = \frac{\cancel{0.7} \cdot n}{\cancel{0.7}}$$

$$180 = n$$

So, 126 credits is 70% of 180 credits.

  b. 475% of what amount is $190?

Translate the sentence into an equation. Write the percent in decimal form.

475%   of $\underbrace{\text{what amount}}$ is $190?

$$4.75 \quad \cdot \quad n \quad = \quad 190$$

To solve, divide both sides by 4.75.

$$\frac{\cancel{4.75} \cdot n}{\cancel{4.75}} = \frac{190}{4.75}$$

$$n = 40$$

So 475% of $40 is $190.

6. Write and solve a percent equation to answer each question.
a. 48 tons is 16% of how many tons?

_____

b. 350% of what amount is 273?

_____

Objective 1   **Estimate answers to percent problems involving 25%.**

For extra help, see Example 1 on page 528 of your text and Section Lecture video for Section 7.3.

Estimate the answer to each question.

  1.   Find 25% of 97 minutes.   **1.** _____

  2.   25% of 43 inches is how many inches?   **2.** _____

  3.   What is 25% of $7823?   **3.** _____

Objective 2   Find 10% and 1% of a number by moving the decimal point.

For extra help, see Examples 2–3 on pages 528–529 of your text and Section Lecture video for Section 7.3 and Exercise Solutions Clip 9.

Find the exact answer to each question by moving the decimal point.

4.   1% of 400 homes is _____ .   4. _____

5.   10% of 4920 televisions is _____ .   5. _____

6.   1% of $98 is _____ .   6. _____

Objective 3   Solve basic percent problems using the percent equation.

For extra help, see Examples 4–6 on pages 530–533 of your text and Section Lecture video for Section 7.3 and Exercise Solutions Clip 21 and 29.

Write and solve an equation to answer each question.

7.   195 calls is what percent of 260 calls?   7. _____

8.   3% of $720 is how much?   8. _____

9.   24 magazines is 6% of what number of magazines?   9. _____

Chapter 7 PERCENT

7.4   Problem Solving with Percent

Learning Objectives
1   Solve percent application problems.
2   Solve problems involving percent of increase or decrease.

Key Terms

Use the vocabulary terms listed below to complete each statement in exercises 1–3.

> **percent of increase or decrease**   **percent proportion**
>
> **percent equation**

1.   The statement $\text{part} = \text{percent} \cdot \text{whole}$ is called the _____.

2.   In a _____ problem, the increase or decrease is expressed as a percent of the original amount.

3.   The statement $\dfrac{\text{part}}{\text{whole}} = \dfrac{\text{percent}}{100}$ is called the _____.

Guided Examples

Review this example for Objective 1:
1.   20% of the students at Lakeville Community College are married. How many of the 3260 students enrolled this year are married?

Step 1 Read the problem. Find the number of married students.

Step 2 Assign a variable. Let n be the number of married students.

Step 3 Write an equation. Use the percent equation.

   $\text{percent} \cdot \text{whole} = \text{part}$

   <u>20%</u>   of   3260   is   <u>how many students?</u>

   $\underset{0.20}{\downarrow} \quad \underset{\cdot}{\downarrow} \quad \underset{3260}{\downarrow} \quad \underset{=}{\downarrow} \qquad \underset{n}{\downarrow}$

Step 4 Solve the equation. Simplify the left side, multiplying 0.20 and 3260.

   $(0.20)(3260) = n$

   $652 = n$

Now Try:
1.   There were 75 points on the first statistics test. Steven's score was 88% correct. How many points did Steven earn?

   _____

Step 5 State the answer. The number of married students is 652.

Step 6 Check the solution. 10% of 3260 is 326. 20% is two times 326, or 652.

2. Louise Trent earns $450 per week and has 14% of this amount withheld for taxes. Find the amount withheld.

Step 1 Read the problem. Find the amount withheld for taxes.

Step 2 Assign a variable. Let n be the amount withheld.

Step 3 Write an equation. Use the percent equation.

percent · whole = part

14%   of   450   is   how much?

$$0.14 \cdot 450 = n$$

Step 4 Solve the equation. Simplify the left side, multiplying 0.14 and 450.

$$(0.14)(450) = n$$
$$63 = n$$

Step 5 State the answer. The amount withheld is $63.

Step 6 Check the solution. 10% of $450 is $45 and 5% of $450 is $22.50. So then $45 + $22.50 = $67.50, so a little less than $67.50 should be withheld.

3. Lin needs 60 credits to graduate from her community college. So far she has earned 51 credits. What percent of the required credits does she have?

Step 1 Read the problem. Determine the credits earned.
    Unknown: percent earned
    Known: earned 51 out of 60 credits.

Step 2 Assign a variable. Let p be the unknown percent.

Step 3 Write an equation.

2. A survey at an intersection found that, of 2450 drivers, 12% were talking on a cell phone. How many drivers in the survey were talking on the phone?

_____

3. Jane eats 1500 calories a day. If she eats 350 calories for breakfast, what percent of her daily calories is her breakfast?

_____

$$\begin{array}{ccccc}
\text{percent} & \cdot & \text{whole} & = & \text{part} \\
\downarrow & & \downarrow & & \downarrow \\
p & \cdot & 60 & = & 51
\end{array}$$

Step 4 Solve the equation.

$$p \cdot 60 = 51$$

$$\frac{p \cdot \cancel{60}}{\cancel{60}} = \frac{51}{60}$$

$$p = 0.85$$

$$0.85 = 85\%$$

Step 5 State the answer. Lin earned 85% of credits needed.

Step 6 Check the solution. Lin earned most of the credits, so the percent should be close to 100%.

4. Janelle had budgeted $250 for new school clothes but ended up spending $390. The amount she spent was what percent of her budget?

Step 1 Read the problem. It is about comparing her budget to the amount spent.
    Unknown: The percent of her budget
    Known: $250 budgeted, $390 spent.

Step 2 Assign a variable. Let p be the unknown percent.

Step 3 Write an equation.

$$\begin{array}{ccccc}
\underbrace{\text{Amount spent}} & \text{is} & \underbrace{\text{what percent}} & \text{of} & \underbrace{\text{amount budgeted?}} \\
\downarrow & \downarrow & \downarrow & \downarrow & \downarrow \\
390 & = & p & \cdot & 250
\end{array}$$

Step 4 Solve the equation.

$$390 = p \cdot 250$$

$$\frac{390}{250} = \frac{p \cdot \cancel{250}}{\cancel{250}}$$

$$1.56 = p$$

$$1.56 = 156\%$$

Step 5 State the answer. Janelle spent 156% of her budget.

Step 6 Check the solution. 100% is $250, and 50% is $125. So $250 + $125 = $350, or 150%, which is close to 156%.

4. Total Fitness Club predicted that 240 new members would join after Christmas. It actually had 396 new members join. The actual number joining is what percent of the predicted number?

_____

Name: Date:
Instructor: Section:

5. David has 4.5% of his earnings deposited into a money market. If this amounts to $146.25 per month, find his monthly earnings.

Step 1 Read the problem. It is about earnings.
   Unknown: monthly salary
   Known: $146.25 is 4.5% of monthly earnings.

Step 2 Assign a variable. Let n = monthly salary.

Step 3 Write an equation.

   4.5%   of   how much   is $146.25

   0.045   · n =   146.25

Step 4 Solve the equation.
   $0.045 \cdot n = 146.25$

   $$\frac{0.045 \cdot n}{0.045} = \frac{146.25}{0.045}$$

   $n = 3250$

Step 5 State the answer. David's monthly earnings is $3250.

Step 6 Check the solution. Round 4.5% to 5%. 10% of $3250 is $325, then 5% is half of $325 or $162.50, which is close to the number given.

5. A multivitamin contains 10 micrograms of vitamin K. If this is 13% of the recommended daily dosage, what is the recommended daily dosage of vitamin K? Round the answer to the nearest whole number.

_____

Review this example for Objective 2:
6. Over the last three years, Calvin's salary has increased from $2700 per month to $3200. What is the percent increase?

Step 1 Read the problem. It is about a salary increase.
   Unknown: percent of increase
   Known: original salary was $2700;
         new salary is $3200.

Step 2 Assign a variable. Let p be the percent of increase.

Step 3 Write an equation. First, subtract $3200 - $2700 to find the amount of increase.
   $3200 - $2700 = $500

   percent of original wage is amount of increase

   p ·   2700   =   500

Now Try:
6. A business increased the number of phone lines from 4 to 9. What is the percent increase?

_____

Step 4 Solve the equation.

$$p \cdot 2700 = 500$$

$$\frac{p \cdot \cancel{2700}}{\cancel{2700}} = \frac{500}{2700}$$

$$p \approx 0.185$$

$$0.185 = 18.5\%$$

Step 5 State the answer. Calvin's salary increased by 18.5%.

Step 6 Check the solution. Round 18.5% to 20%. Then 10% of $2700 is $270, and 20% is two times 10%, or $540, which is close to the number given.

7. During the holiday season, average daily attendance at a health club fell from 495 members to 230 members. What was the percent decrease?

Step 1 Read the problem. It is about decrease in attendance.

    Unknown: percent of decrease
    Known: original attendance: 495
               new attendance: 230.

Step 2 Assign a variable. Let p be the percent of decrease.

Step 3 Write an equation. First subtract 230 from 495 to find the amount of decrease.

    $495 - 230 = 265$

$$\underbrace{\text{percent}}\ \text{of}\ \underbrace{\begin{array}{c}\text{original}\\\text{attendance}\end{array}}\ \text{is}\ \underbrace{\begin{array}{c}\text{amount}\\\text{of decrease}\end{array}}$$

$$\begin{array}{ccccc}\downarrow & \downarrow & \downarrow & \downarrow & \downarrow \\ p & \cdot & 495 & = & 265\end{array}$$

Step 4 Solve the equation.

$$p \cdot 495 = 265$$

$$\frac{p \cdot \cancel{495}}{\cancel{495}} = \frac{265}{495}$$

$$p \approx 0.535$$

$$0.535 = 53.5\%$$

Step 5 State the answer. The attendance decreased by 53.5%.

Step 6 Check the solution. 50% is half of 495, or 247.5, so 53.5% is a reasonable solution.

7. The earnings per share of Amy's Cosmetic Company decreased from $1.20 to $0.86 in the last year. Find the percent of decrease.

_____

Objective 1   Solve percent application problems.

For extra help, see Examples 1–5 on pages 540–543 of your text and Section Lecture video for Section 7.4 and Exercise Solutions Clip 11 and 13.

Use the six problem-solving steps to answer each question. Round percent answers to the nearest tenth of a percent.

1.   Members who are between 25 and 45 years of age make up 92% of the total membership of an organization. If there are 850 total members in the organization, find the number of members in the 25 to 45 age group.

1. _____

2.   Payroll deductions are 35% of Jason's gross pay. If his deductions total $350, what is his gross pay?

2. _____

3.   Vera's Antique Shoppe says that of its 5100 items in stock, 4233 are just plain junk, while the rest are antiques. What percent of the number of items in stock is antiques?

3. _____

Objective 2   Solve problems involving percent of increase or decrease.

For extra help, see Examples 6–7 on pages 544–545 of your text and Section Lecture video for Section 7.4 and Exercise Solutions Clip 27.

Use the six problem-solving steps to find the percent increase or decrease. Round your answers to the nearest tenth of a percent.

4.   Students at Withrow's College were charged $1560   4. _____
      for tuition this semester. If the tuition was $1480 last
      semester, find the percent of increase.

5.   Tomika's part-time work schedule has been reduced   5. _____
      to 20 hours per week. She had been working 28
      hours per week. What is the percent decrease?

6.   During a sale, the price of a futon was cut from   6. _____
      $1250 to $999. Find the percent of decrease in price.

Chapter 7 PERCENT

7.5   Consumer Applications: Sales Tax, Tips, Discounts, and Simple Interest

Learning Objectives
1   Find sales tax and total cost.
2   Estimate and calculate restaurant tips.
3   Find the discount and sale price.
4   Calculate simple interest and the total amount due on a loan.

Key Terms

Use the vocabulary terms listed below to complete each statement in exercises 1–8.

sales tax	tax rate	discount
interest	principal	interest rate
simple interest	interest formula	

1.   The charge for money borrowed or loaned, expressed as a percent, is called
      _____.

2.   A fee paid for borrowing or lending money is called _____.

3.   The formula $I = p \cdot r \cdot t$ is the _____.

4.   Use the formula $I = p \cdot r \cdot t$ to compute the amount of _____
      due on a loan.

5.   The amount of money borrowed or loaned is called the _____.

6.   The percent of the total sales charged as tax is called the _____.

7.   The percent of the original price that is deducted from the original price is called
      the _____.

8.   The percent used when calculating the amount of tax is called the
      _____.

Guided Examples

Review these examples for Objective 1:

1.   A television sells for $750 plus 8% sales tax.
      Find the price of the TV including sales tax.

      Step 1 Read the problem. It asks for the tax and
      the price of the TV.

      Step 2 Assign a variable. Let *n* be the amount of

Now Try:

1.   A cell phone is $89. The sales
      tax is 7%. What is the tax and
      total?

      _____

tax.

Step 3 Write an equation. Use the sales tax equation.

$$\underbrace{\text{tax rate}} \cdot \underbrace{\text{cost of item}} = \underbrace{\text{sales tax}}$$
$$0.08 \quad \cdot \quad 750 \quad = \quad n$$

Step 4 Solve.

$$(0.08)(750) = n$$
$$60 = n$$

Add the sales tax to the cost of the TV to find the total cost.

$$\underbrace{\text{cost of item}} + \underbrace{\text{sales tax}} = \underbrace{\text{total cost}}$$
$$\$750 \quad + \quad \$60 \quad = \quad \$810$$

Step 5 State the answer. The tax is $60 and the total cost of the TV, including tax, is $810.

Step 6 Check. Use estimation to check that the amount of sales tax is reasonable. Round 8% to 10%. Then 10% of $750 is $75. It is reasonable.

2. A gold bracelet costs $1300 not including a sales tax of $71.50. Find the sales tax rate.

Step 1 Read the problem. It asks for the tax rate.

Step 2 Assign a variable. Let p be the tax rate (the percent).

Step 3 Write an equation.

$$\underbrace{\text{tax rate}} \cdot \underbrace{\text{cost of item}} = \underbrace{\text{sales tax}}$$
$$p \quad \cdot \quad 1300 \quad = \quad 71.50$$

Step 4 Solve.

$$p \cdot 1300 = 71.50$$
$$\frac{p \cdot \cancel{1300}}{\cancel{1300}} = \frac{71.50}{1300}$$
$$p = 0.055$$
$$0.055 = 5.5\%$$

Step 5 State the answer. The sales tax rate is 5.5%.

Step 6 Check. Use estimation to check that the answer is reasonable. 1% of $1300 is $13. Round 5.5% to 6%. Then 6% is six times $13 or $78. The solution is reasonable.

2. The rate for a hotel room is $124. The tax is $14.88. Find the tax rate.

_____

Review these examples for Objective 2:

3. First estimate each tip. Then calculate the exact answer.

    a. Find a 15% tip for a restaurant bill of $87.22.

    Estimate: Round $87.22 to $90. Then 10% of $90 is $9, and 5% is half of $9, or $4.50. So the estimate is $9 + $4.50 = $13.50.

    Exact: Use the percent equation. Write 15% as a decimal. The bill for food and beverages is the whole and the tip is the part.

percent	·	whole	=	part
↓		↓		↓
15%	·	$87.22	=	n

    $$(0.15)(87.22) = n$$
    $$13.083 = n$$

    Round $13.083 to $13.08 (nearest cent), which is close to the estimate of $13.50.

    b. Find a 20% tip for a restaurant bill of $87.22.

    Estimate: Round $87.22 to $90. Then 10% of $90 is $9, and 20% is two times $9 is $18.

    Exact: Use the percent equation. Write 20% as a decimal. The bill for food and beverages is the whole and the tip is the part.

percent	·	whole	=	part
↓		↓		↓
20%	·	$87.22	=	n

    $$(0.20)(87.22) = n$$
    $$17.444 = n$$

    Round $17.444 to $17.44 (nearest cent), which is close to the estimate of $18.

Review this example for Objective 3:

4. Mike Lee can purchase a new car at 8% below window sticker price. Find the sale price on a car with a window sticker price of $17,600.

    Step 1 Read the problem. The problem asks for the price of the car after a discount of 8%.

    Step 2 Work out a plan. The problem is solved in two steps. First, find the amount of the discount, that is, the amount that will be "taken off" (subtracted), by multiplying the original price ($17,600) by the rate of the discount (8%).

Now Try:

3. First estimate each tip. Then calculate the exact answer.
    a. Find a 15% tip for a restaurant bill of $23.75.

    _____

    b. Find a 20% tip for a restaurant bill of $23.75.

    _____

Now Try:

4. A hard-cover book with an original price of $24.95 is on sale at 60% off. Find the sale price of the book.

    _____

The second step is to subtract the amount of discount from the original price. This gives the sale price, which is what Mike will actually pay for the car.

Step 3 Estimate a reasonable answer. Round the original price from $17,600 to $20,000, and the rate of discount from 8% to 10%. Since 10% is equivalent to $\frac{1}{10}$, the estimated discount is $20,000 \div 10 = 2000, so the estimated sale price is $20,000 - $2000 = $18,000.

Step 4 Solve the problem. First find the exact amount of the discount.

amount of discount = rate of discount · original price
$$a = (8\%)(\$17,600)$$
$$a = (0.08)(\$17,600)$$
$$a = \$1408$$

Now, find the sale price of the car by subtracting the amount of the discount ($1408) from the original price.

sale price = original price − amount of discount
$$= \$17,600 - \$1408$$
$$= \$16,192$$

Step 5 State the answer. The sale price of the car is $16,192.

Step 6 Check. The exact answer, $16,192, is close to the estimate of $18,000.

Review these examples for Objective 4:
5. Find the simple interest and total amount due on $4000 at 4% for 1 year.

The amount borrowed, or principal (*p*), is $4000. The interest rate (*r*), is 4%, which is 0.04 as a decimal, and the time of the loan (*t*) is 1 year. Use the formula.
$$I = p \cdot r \cdot t$$
$$I = (4000)(0.04)(1)$$
$$I = \$160$$
The interest is $160.
Now add the principal and the interest to find the

Now Try:
5. Find the interest and total amount due on $80 at 5% for 1 year.

_____

total amount due.

amount due = principal + interest

$$= \$4000 + \$160$$

$$= \$4160$$

The total amount due is $4160.

6. Find the simple interest and total amount due on $5600 at 5% for two and a half years.

The principal (p) is $5600. The rate ($r$) is 5%, or 0.05 as a decimal, and the time (t) is $2\frac{1}{2}$ or 2.5 years. Use the formula.

$$I = p \cdot r \cdot t$$

$$I = (5600)(0.05)(2.5)$$

$$I = \$700$$

The interest is $700.
Now add the principal and the interest to find the total amount due.

amount due = principal + interest

$$= \$5600 + \$700$$

$$= \$6300$$

The total amount due is $6300.

7. Find the simple interest and total amount due on $720 at $3\frac{1}{2}$% for 8 months.

The principal is $720. The rate is $3\frac{1}{2}$% or 0.035 as a decimal, and the time is $\frac{8}{12}$ of a year. Use the formula $I = p \cdot r \cdot t$.

$$I = (720)(0.035)\left(\frac{8}{12}\right)$$

$$= 25.2\left(\frac{2}{3}\right)$$

$$= 16.80$$

The interest is $16.80.
The total amount due is
$720 + $16.80 = $736.80.

6. Find the simple interest and total amount due on $620 at 16% for one and a quarter years.

_____

7. Find the simple interest and total amount due on $840 at $8\frac{1}{2}$% for 5 months.

_____

Objective 1   Find sales tax and total cost.

For extra help, see Examples 1–2 on pages 550–551 of your text and Section Lecture video for Section 7.5 and Exercise Solutions Clip 7.

Find the amount of sales tax and the total cost. Round answers to the nearest cent, if necessary.

	Amount of sale	**Tax Rate**
1.	$350	6.5%

1. Tax_____

   Total_____

Find the sales tax rate. Round answers to the hundredth, if necessary.

	Amount of sale	**Amount of Tax**
2.	$450	$36

2. _____

3.	$215	$10.75

3. _____

Objective 2   Estimate and calculate restaurant tips.

For extra help, see Example 3 on pages 551–552 of your text and Section Lecture video for Section 7.5 and Exercise Solutions Clip 49.

For each restaurant bill, estimate a 15% tip and a 20% tip. Then find the exact amounts for a 15% tip and a 20% tip. Round exact amounts to the nearest cent if necessary.

4. $43.16

4. 15% estimate _____

   15% exact_____

   20% estimate _____

   20% exact_____

5. $63.85

5. 15% estimate _____

   15% exact_____

   20% estimate _____

   20% exact_____

6. $72.81

6. 15% estimate _____

15% exact_____

20% estimate _____

20% exact_____

Objective 3   Find the discount and sale price.

For extra help, see Example 4 on page 553 of your text and Section Lecture video for Section 7.5.

Find the amount of discount and the amount paid after the discount. Round money answers to the nearest cent, if necessary.

	Original price	**Rate of Discount**
7.	$200	15%

7. Discount_____

Amount paid _____

Solve each application problem. Round money answers to the nearest cent, if necessary.

8.   A "Super 35% Off Sale" begins today. What is the price of a hair dryer normally priced at $15?

8. _____

9.   Geishe's Shoes sells shoes at 33% off the regular price. Find the price of a pair of shoes normally priced at $54, after the discount is given.

9. _____

Objective 4   Calculate simple interest and the total amount due on a loan.

For extra help, see Examples 5–7 on pages 554–555 of your text and Section Lecture video for Section 7.5 and Exercise Solutions Clip 35 and 47.

Find the simple interest and the total amount due. Round to the nearest cent, if necessary.

	Principal	**Rate**	**Time in Years**
10.	$5280	9%	1
11.	$780	10%	$2\frac{1}{2}$
12.	$14,400	7%	7 months

10. _____

_____

11. _____

_____

12. _____

_____

Chapter 8 MEASUREMENT

8.1   Problem Solving with U.S. Measurement Units

Learning Objectives
1   Learn the basic U.S. measurement units.
2   Convert among U.S. measurement units using multiplication or division.
3   Convert among measurement units using unit fractions.
4   Solve application problems using U.S. measurement units.

Key Terms

Use the vocabulary terms listed below to complete each statement in exercises 1–3.

      U.S. measurement units   **unit fractions**   **metric system**

1.   The _____ is based on units of ten.

2.   _____ are used to convert among different measurements.

3.   _____ include inches, feet, quarts, and pounds.

Guided Examples

Review these examples for Objective 1:

1.   Memorize the U.S. measurement conversions from the text. Then fill in the blanks.

    a. 2 c = _____ pt

Answer: 1 pt

    b. 1 mi = _____ ft

Answer: 5280 ft

Now Try:

1.   Memorize the U.S. measurement conversions from the text. Then fill in the blanks.

    a. 16 oz = _____ lb

    _____

    b. 1 gal = _____ qt

    _____

Review these examples for Objective 2:

2.   Convert each measurement.

    a. 16 yd to feet

You are converting from a larger unit to a smaller unit, so multiply.
Because 1 yd = 3 ft, multiply by 3.
    $16 \text{ yd} = 16 \cdot 3 = 48 \text{ ft}$

Now Try:

2.   Convert each measurement.

    a. $17\frac{1}{2}$ ft to inches

    _____

b. 6 T to pounds

You are converting from a larger unit to a smaller unit, so multiply.
Because 1 T = 2000 lb, multiply by 2000.

$$6 \text{ T} = 6 \cdot 2000 = 12{,}000 \text{ lb}$$

c. 8 pt to quarts

You are converting from a smaller unit to a larger unit, so divide.
Because 2 pt = 1 qt, divide by 2.

$$8 \text{ pt} = \frac{8}{2} = 4 \text{ qt}$$

d. 3960 ft to miles

You are converting from a smaller unit to a larger unit, so divide.
Because 5280 ft = 1 mi, divide by 5280.

$$3960 \text{ ft} = \frac{3960}{5280} = \frac{3}{4} \text{ mi}$$

b. $3\frac{1}{2}$ lb to ounces

_____

c. 75 sec to minutes

_____

d. 380 min to hours

_____

Review these examples for Objective 3:

3.

  a. Convert 64 oz to pounds.

  Use a unit fraction with pounds (the unit for your answer) in the numerator, and ounces (the unit being changed) in the denominator. Because 1 lb = 16 oz, the necessary unit fraction is

$$\frac{1 \text{ lb}}{16 \text{ oz}} \quad \begin{matrix} \leftarrow \text{ Unit for your answer is pounds.} \\ \leftarrow \text{ Unit being changed is ounces.} \end{matrix}$$

  Next, multiply 64 oz times this unit fraction.

  Write 64 oz as the fraction $\frac{64 \text{ oz}}{1}$ and divide out common units and factors wherever possible.

$$\frac{64 \text{ oz}}{1} \cdot \frac{1 \text{ lb}}{16 \text{ oz}} = \frac{\overset{4}{\cancel{64}} \ \cancel{oz}}{1} \cdot \frac{1 \text{ lb}}{\underset{1}{\cancel{16}} \ \cancel{oz}} = \frac{4 \cdot 1 \text{ lb}}{1} = 4 \text{ lb}$$

  b. Convert 8 lb to ounces.

  Select the correct unit fraction to change 8 lb to ounces.

$$\frac{16 \text{ oz}}{1 \text{ lb}} \quad \begin{matrix} \leftarrow \text{ Unit for your answer is ounces.} \\ \leftarrow \text{ Unit being changed is pounds.} \end{matrix}$$

  Multiply 8 lb times the unit fraction.

$$\frac{8 \text{ lb}}{1} \cdot \frac{16 \text{ oz}}{1 \text{ lb}} = \frac{8 \ \cancel{lb}}{1} \cdot \frac{16 \text{ oz}}{1 \ \cancel{lb}} = \frac{8 \cdot 16 \text{ oz}}{1} = 128 \text{ oz}$$

Now Try:

3.

  a. Convert 12 yd to inches.

  _____

  b. Convert 4 in. to feet.

  _____

4. Convert using unit fractions.

a. Convert 6 qt to gallons.

First select the correct unit fraction.

$\dfrac{1\ gal}{4\ qt}$ ← Unit for your answer is gallons.
← Unit being changed is quarts.

Next multiply.

$$\frac{6\ qt}{1}\cdot\frac{1\ gal}{4\ qt}=\frac{\overset{3}{\cancel{6}}\ \cancel{qt}}{1}\cdot\frac{1\ gal}{\underset{2}{\cancel{4}}\ \cancel{qt}}=\frac{3}{2}\ gal=1\frac{1}{2}\ gal$$

b. Convert $6\dfrac{1}{2}$ lb to ounces.

Write $6\dfrac{1}{2}$ lb as an improper fraction.

$$\frac{6\frac{1}{2}\ \cancel{lb}}{1}\cdot\frac{16\ oz}{1\ \cancel{lb}}=\frac{13}{2}\cdot\frac{16}{1}\ oz$$

$$=\frac{13}{\underset{1}{\cancel{2}}}\cdot\frac{\overset{8}{\cancel{16}}}{1}\ oz$$

$$=104\ oz$$

c. Convert 340 hr to days.

$$\frac{\overset{85}{\cancel{340}}\ \cancel{hr}}{1}\cdot\frac{1\ day}{\underset{6}{\cancel{24}}\ \cancel{hr}}=\frac{85}{6}\ day=14\frac{1}{6}\ day$$

5. Sometimes you may need to use two or three unit fractions to complete a conversion.

a. Convert 110 pt to gallons.

Use the unit fraction $\dfrac{1\ qt}{2\ pt}$ to change pints to

quarts and the unit fraction $\dfrac{1\ gal}{4\ qt}$ to change

quarts to gallons. Notice how all the units divide out except gallons, which is the unit you want in the answer.

$$\frac{110\ \cancel{pt}}{1}\cdot\frac{1\ \cancel{qt}}{2\ \cancel{pt}}\cdot\frac{1gal}{4\ \cancel{qt}}=\frac{110}{8}gal=\frac{110\div2}{8\div2}gal=13\frac{3}{4}gal$$

13.75 gal is equivalent to $13\dfrac{3}{4}$ gal.

4. Convert using unit fractions.

a. Convert 4 mi to feet.

_____

b. Convert 3000 lb to tons.

_____

c. Convert $3\dfrac{1}{4}$ gal to quarts.

_____

5. Sometimes you may need to use two or three unit fractions to complete a conversion.

a. Convert 12 cups to gallons.

_____

b. Convert 3 miles to inches.

Use three unit fractions. The first one changes from miles to yards, the next one changes yards to feet, and the last one changes feet to inches. All the units divide out except inches, which is the unit you want in your answer.

$$\frac{3\,\text{mi}}{1} \cdot \frac{1760\,\text{yd}}{1\,\text{mi}} \cdot \frac{3\,\text{ft}}{1\,\text{yd}} \cdot \frac{12\,\text{in.}}{1\,\text{ft}} = 190,080 \text{ inches}$$

b. Convert 5 days to minutes.

_____

Review these examples for Objective 4:

6. Answer each application problem.

a. Lee paid $4.65 for 14 oz of honey baked ham. What is the price per pound, to the nearest cent?

Step 1 Read the problem. The problem asks for the price per pound of the ham.

Step 2 Work out a plan. The weight of the ham is given in ounces, but the answer must be cost per pound. Convert ounces to pounds. The word *per* indicates division. You need to divide the cost by the number of pounds.

Step 3 Estimate a reasonable answer. To estimate, round $4.65 to $5, and round 14 ounces to 16 ounces. Then there are 16 oz in a pound, so there is about 1 pound. Finally, $5 \div 1 = $5 per pound is our estimate.

Step 4 Solve the problem. Use a unit fraction to convert 14 oz to pounds.

$$\frac{\overset{7}{\cancel{14}}\,\cancel{oz}}{1} \cdot \frac{1\,\text{lb}}{\underset{8}{\cancel{16}}\,\cancel{oz}} = \frac{7}{8}\,\text{lb} = 0.875 \text{ lb}$$

Then divide to find the cost per pound.

$$\frac{\$4.65}{0.875} \approx \$5.31$$

Step 5 State the answer. The ham is $5.31 per pound (nearest cent).

Step 6 Check your work. The exact answer, $5.31 is close to the estimate of $5.

Now Try:

6. Answer each application problem.
a. Clarissa paid $1.79 for 4.5 oz of nuts. What is the cost per pound, to the nearest cent?

_____

b. At a preschool, each of 20 children drinks about $\frac{3}{4}$ c of juice with their snack each day. The school is open 5 days a week. How many quarts of juice are needed for 1 week of snacks?

Step 1 Read the problem. The problem asks for the number of quarts of juice needed for 1 week.

Step 2 Work out a plan. Multiply to find the number of cups of juice for one week. Then convert cups to quarts.

Step 3 Estimate a reasonable answer. To estimate, round $\frac{3}{4}$ cups to 1 cup. Then 1 cup times 20 children times 5 days is 100 cups. There are 4 cups in a quart. So, 100 cups $\div 4 = 25$ quarts is our estimate.

Step 4 Solve the problem. First multiply. Then use unit fractions to convert.

$$\frac{3}{4} \cdot 20 \cdot 5 = \frac{3}{\cancel{4}} \cdot \frac{\overset{5}{\cancel{20}}}{1} \cdot \frac{5}{1} = 75 \text{ cups}$$

$$\frac{75 \ \cancel{\text{cups}}}{1} \cdot \frac{1 \ \cancel{\text{pt}}}{2 \ \cancel{\text{cups}}} \cdot \frac{1 \ \text{qt}}{2 \ \cancel{\text{pt}}} = \frac{75}{4} \text{ qt} = 18\frac{3}{4} \text{ qt}$$

Step 5 State the answer. The preschool needs $18\frac{3}{4}$ qt of juice for 1 week.

Step 6 Check your work. The exact answer of $18\frac{3}{4}$ qt is close to our estimate of 25 qt.

b. Sweet Suzie's Shop makes 49,000 lb of fudge each week. How many tons of fudge are produced each day of the 7-day week?

_____

Objective 1   Learn the basic U.S. measurement units.

For extra help, see Example 1 on page 578 of your text and Section Lecture video for Section 8.1.

Fill in the blanks.

1.   1 T = _____ lb

1. _____

2.   _____ qt = 1 gal

2. _____

3.   1 c = _____ fl oz   **3.** _____

Objective 2   **Convert among U.S. measurement units using multiplication or division.**

For extra help, see Example 2 on page 579 of your text and Section Lecture video for Section 8.1 and Exercise Solutions Clip 11.

Convert each measurement using multiplication or division.

4.   12 ft to yards   **4.** _____

5.   40 pt to gallons   **5.** _____

6.   30 in to yards   **6.** _____

Objective 3   **Convert among measurement units using unit fractions.**

For extra help, see Examples 3–5 on pages 580–582 of your text and Section Lecture video for Section 8.1 and Exercise Solutions Clip 39 and 41.

Convert each measurement using unit fractions.

7.   28 pt to gallons   **7.** _____

8.   38 c to pints   **8.** _____

9.   60 oz to pounds   **9.** _____

Objective 4   Solve application problems using U.S. measurement units.

For extra help, see Example 6 on pages 582–583 of your text and Section Lecture video for Section 8.1 and Exercise Solutions Clip 57.

Solve each application problem using the six problem-solving steps.

10.   Tony paid $2.84 for 6.5 oz of fudge. What is the cost   **10.** _____
        per pound, to the nearest cent?

11.   At an office, each of 15 workers drinks about $1\frac{3}{4}$ c   **11.** _____
        of coffee with each day. The office is open 4 days a
        week. How many quarts of coffee are needed for a
        week?

Chapter 8 MEASUREMENT

8.2   The Metric System—Length

Learning Objectives
1   Learn the basic metric units of length.
2   Use unit fractions to convert among metric units.
3   Move the decimal point to convert among metric units.

Key Terms

Use the vocabulary terms listed below to complete each statement in exercises 1–3.

   meter   **prefix**   **metric conversion line**

1.   Attaching a _____ such as "kilo-" or "milli" to the words "meter", "liter", or "gram" gives the names of larger or smaller units.

2.   A line showing the various metric measurement prefixes and their size relationship to each other is called a _____.

3.   The basic unit of length in the metric system is the _____.

Guided Examples

Review these examples for Objective 1:

1.   Write the most reasonable metric unit in each blank. Choose from km, m, cm, and mm.

   a. The man's foot is 28 _____ .

28 cm because cm are used instead of inches. 28 cm is about 11 inches.

   b. The living room is 4 _____ .

4 m because m are used instead of feet. 4 m is about 13 feet.

   c. Sam rode his bike 8 _____ on the trail.

8 km because km are used instead of miles. 8 km is about 5 miles.

Now Try:

1.   Write the most reasonable metric unit in each blank. Choose from km, m, cm, and mm.

   a. A strand of hair is 1 ____ wide.

   _____

   b. The truck sped down the highway at 120 _____ per hour.

   _____

   c. The width of a piece of paper is 21.6 _____ .

   _____

Name: _____ Date: _____
Instructor: _____ Section: _____

Review these examples for Objective 2:

2. Convert each measurement using unit fractions.

   a. 8 m to cm

   Put the unit for the answer (cm) in the numerator of the unit fraction; put the unit you want to change (m) in the denominator.

   $$\frac{100 \text{ cm}}{1 \text{ m}} \leftarrow \text{Unit for answer} \\ \leftarrow \text{Unit being changed}$$

   Multiply. Divide out common units where possible.

   $$8\text{m} \cdot \frac{100\text{cm}}{1\text{m}} = \frac{8\,\cancel{\text{m}}}{1} \cdot \frac{100\text{cm}}{1\,\cancel{\text{m}}} = \frac{8 \cdot 100\text{cm}}{1} = 800\text{cm}$$

   8 m = 800 cm

   b. 13.5 mm to cm

   Multiply by a unit fraction that allows you to divide out millimeters.

   $$\frac{13.5\,\cancel{\text{mm}}}{1} \cdot \frac{1 \text{ cm}}{10\,\cancel{\text{mm}}} = \frac{13.5 \text{ cm}}{10} = 1.35 \text{ mm}$$

   13.5 mm = 1.35 cm
   There are 10 mm in a cm, so 13.5 mm will be a smaller part of a cm.

Now Try:

2. Convert each measurement using unit fractions.
   a. 5400 m to km

   _____

   b. 7.6 cm to mm

   _____

				$\frac{1}{10}$	$\frac{1}{100}$	$\frac{1}{1000}$
1000	100	10	1			
km	hm	dam	m	dm	cm	mm

Review these examples for Objective 3:

3. Use the metric conversion line to make the following conversions.

   a. 62.892 km to m

   Find km on the metric conversion line. To get m, you move three places to the right. So move the decimal point in 62.892 three places to the right.
   62.892 km = 62,892 m

   b. 47.6 cm to m

   Find cm on the conversion line. To get m, move two places to the left. So move the decimal point two places to the left.
   47.6 cm = 0.476 m

   c. 52.8 mm to cm

   From mm to cm is one place to the left.
   52.8 mm = 5.28 cm

Now Try:

3. Use the metric conversion line to make the following conversions.
   a. 67.5 cm to mm

   _____

   b. 986.5 m to km

   _____

   c. 4.31 m to cm

   _____

4. Convert using the metric conversion line.

    a. 1.79 m to mm

Moving from m to mm is going three places to the right. In order to move the decimal point in 1.79 three places to the right, you must add a 0 as a place holder.
    1.790

1.79 m = 1790 mm

    b. 50 cm to m

From cm to m is two places to the left. The decimal point in 50 starts at the far right side because 50 is a whole number. Then move it two places to the left.

50 cm = 0.50 m and 0.50 m is equivalent to 0.5 m.

    c. 19 m to km

From m to km is three places to the left. The decimal point in 19 starts at the far right side. In order to move it three places to the left, you must write in one zero as a placeholder.

19 m = 0.019 km

4. Convert using the metric conversion line.

    a. 23 m to mm

    _____

    b. 4.2 cm to m

    _____

    c. 950 m to km

    _____

Objective 1   Learn the basic metric units of length.

For extra help, see Example 1 on page 589 of your text and Section Lecture video for Section 8.2 and Exercise Solutions Clip 13, 15, 17, and 19.

Choose the most reasonable metric unit. Choose from **km, m, cm,** *or* **mm**.

    1.   the width of a twin bed   **1.** _____

    2.   the thickness of a dime   **2.** _____

    3.   the distance driven in 2 hours   **3.** _____

Name: Date:

Instructor: Section:

Objective 2   Use unit fractions to convert among metric units.

For extra help, see Example 2 on page 590 of your text and Section Lecture video for Section 8.2.

Convert each measurement using unit fractions.

   4.   25.87 m to centimeters   **4.** _____

   5.   450 m to kilometers   **5.** _____

   6.   140 millimeters to meters   **6.** _____

Objective 3   Move the decimal point to convert among metric units.

For extra help, see Examples 3–4 on pages 591–592 of your text and Section Lecture video for Section 8.2 and Exercise Solutions Clip 29, 31, and 35.

Convert each measurement using the metric conversion line.

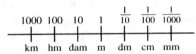

   7.   1.94 cm to millimeters   **7.** _____

   8.   10.35 km to meters   **8.** _____

   9.   3.5 cm to kilometers   **9.** _____

Chapter 8 MEASUREMENT

8.3 The Metric System—Capacity and Weight (Mass)

Learning Objectives
1   Learn the basic metric units of capacity.
2   Convert among metric capacity units.
3   Learn the basic metric units of weight (mass).
4   Convert among metric weight (mass) units.
5   Distinguish among basic metric units of length, capacity, and weight (mass).

Key Terms

Use the vocabulary terms listed below to complete each statement in exercises 1–2.

  liter   gram

1.   The basic unit of weight (mass) in the metric system is the _____.

2.   The basic unit of capacity in the metric system is the _____.

Guided Examples

Review these examples for Objective 1:

1.   Write the most reasonable metric unit in each blank. Choose from L and mL.

   a. The bottle of sunscreen held 118 _____.

   118 mL because 118 L would be about 118 qts, which is too much.

   b. The soda bottle has 2 _____.

   2 L because 2 mL would be less than a teaspoon.

Now Try:

1.   Write the most reasonable metric unit in each blank. Choose from L and mL.
   a. Ty added 5 _____ to the tea.

   _____

   b. The water tank holds 100 _____.

   _____

Review these examples for Objective 2:

2.   Convert using the metric conversion line or unit fractions.
   a. 5.4 L to mL

Using the metric conversion line:
From L to mL is three places to the right.
   5.4 L = 5400 mL

Using unit fractions:
Multiply by a unit fraction that allows you to divide out liters.

$$\frac{5.4 \ \cancel{L}}{1} \cdot \frac{1000 \text{ mL}}{1 \ \cancel{L}} = 5400 \text{ mL}$$

Now Try:

2.   Convert using the metric conversion line or unit fractions.
   a. 973 mL to L

   _____

b. 92 mL to L

Using the metric conversion line:
From mL to L is three places to the left.
    92 mL = 0.092 L

Using unit fractions:
Multiply by a unit fraction that allows you to divide out mL.

$$\frac{9.2 \text{ mL}}{1} \cdot \frac{1 \text{ L}}{1000 \text{ mL}} = 0.092 \text{ L}$$

b. 3.85 L to mL

_____

Review these examples for Objective 3:

3. Write the most reasonable metric unit in each blank. Choose from kg, g, and mg.

a. The watermelon weighed 18 _____ .

18 kg because kilograms are used instead of pounds.
18 kg is about 40 pounds.

b. The vitamin tablet was 400 _____ .

400 mg because 400 g would be more than two hamburgers, which is too much.

c. A birthday card weighs 8 _____ .

8 grams because 8 kg would be too heavy.

Now Try:

3. Write the most reasonable metric unit in each blank. Choose from kg, g, and mg.
a. A fashion model weighs 48 _____ .

_____

b. Jacob's football weighs 590 _____ .

_____

c. The pencil lead weighs 3 _____ .

_____

Review these examples for Objective 4:

4. Convert using the metric conversion line or unit fractions.

a. 26 mg to g

Using the metric conversion line:
From mg to g is three places to the left.
    26 mg = 0.026 g

Using unit fractions:
Multiply by a unit fraction that allows you to divide out mg.

$$\frac{26 \text{ mg}}{1} \cdot \frac{1 \text{ g}}{1000 \text{ mg}} = 0.026 \text{ g}$$

Now Try:

4. Convert using the metric conversion line or unit fractions.

a. 3.72 g to mg

_____

---

b. 97.34 kg to g

Using the metric conversion line:
From kg to g is three places to the right.
    97.34 kg = 97,340 g

Using unit fractions:
Multiply by a unit fraction that allows you to divide out kg.

$$\frac{97.34 \ \cancel{kg}}{1} \cdot \frac{1000 \ g}{1 \ \cancel{kg}} = 97,340 \ g$$

b. 84 g to kg

_____

Review these examples for Objective 5:

5. First decide which type of unit is needed: length, capacity, or weight. Then write the most appropriate metric unit in the blank. Choose from km, m, cm, mm, L, mL, kg, g, and mg.

a. The tree is 9 _____ high.

Use length units because of the word high.
The tree is 9 m high.

b. A ten-carat diamond weighs 2 _____ .

Use weight units because of the word weigh.
The ten-carat diamond weighs 2 g.

c. The milk container has 3.78 _____ .

Use capacity units because milk is a liquid.
The milk container has 3.78 L.

Now Try:

5. First decide which type of unit is needed: length, capacity, or weight. Then write the most appropriate metric unit in the blank. Choose from km, m, cm, mm, L, mL, kg, g, and mg.

a. The moisturizer jar has 125 _____ .

_____

b. This is a 1.19 _____ box of breakfast cereal.

_____

c. The boy walked 400 _____ to the bus stop.

_____

Objective 1   Learn the basic metric units of capacity.

For extra help, see Example 1 on page 596 of your text and Section Lecture video for Section 8.3 and Exercise Solutions Clip 11.

Choose the most reasonable metric unit. Choose from **L,** *or* **ml**.

1.   the amount of soda in a can

1. _____

2.   the amount of orange juice in a large bottle

2. _____

3.   the amount of cough syrup in one dose

3. _____

Name:

Date:

Instructor:

Section:

Objective 2   Convert among metric capacity units.

For extra help, see Example 2 on pages 596–597 of your text and Section Lecture video for Section 8.3.

Convert each measurement. Use unit fractions or the metric conversion line.

4. 2.5 L to milliliters

4. _____

5. 836 kL to liters

5. _____

6. 7863 mL to liters

6. _____

Objective 3   Learn the basic metric units of weight (mass).

For extra help, see Example 3 on page 598 of your text and Section Lecture video for Section 8.3 and Exercise Solutions Clip 7 and 13.

*Choose the most reasonable metric unit. Choose from **kg**, **g**, *or* **mg**.*

7. the weight (mass) of a vitamin pill

7. _____

8. the weight (mass) of a car

8. _____

9. the weight (mass) of an egg

9. _____

Objective 4   Convert among metric weight (mass) units.

For extra help, see Example 4 on page 599 of your text and Section Lecture video for Section 8.3 and Exercise Solutions Clip 39, 41, and 43.

Convert each measurement. Use unit fractions or the metric conversion line.

10. 27,000 g to kilograms

10. _____

11. 0.76 kg to grams

11. _____

12.   4.7 g to milligrams   12. _____

Objective 5   Distinguish among basic metric units of length, capacity, and weight (mass).

For extra help, see Example 5 on page 600 of your text and Section Lecture video for Section 8.3 and Exercise Solutions Clip 57 and 61.

*Choose the most reasonable metric unit. Choose from **km, m, cm, mm, mL, L, kg, g,** or **mg**.*

13.   Buy a 5 _____ bottle of water.   13. _____

14.   The piece of wood weighs 5 _____ .   14. _____

15.   A paperclip is 3 _____ long.   15. _____

Chapter 8 MEASUREMENT

8.4   Problem Solving with Metric Measurement

Learning Objectives
1   Solve application problems involving metric measurements.

Key Terms

Use the vocabulary terms listed below to complete each statement in exercises 1–3.

        meter   **liter**   **gram**

1.   A _____ is the weight of 1 mL of water.

2.   A _____ is a little longer than a yard.

3.   A _____ is a little more than one quart.

Guided Examples

Review these examples for Objective 1:

1. Colby cheese is on sale at $9.89 per kilogram. Nelson bought 750 g of the cheese. How much did he pay, to the nearest cent?

   Step 1 Read the problem. The problem asks for the cost of 750 g of cheese.

   Step 2 Work out a plan. The price is $9.89 per kilogram, but the amount Nelson bought is given in grams. Convert grams to kilograms. Then multiply the weight by the cost per kilogram.

   Step 3 Estimate a reasonable answer. Round the cost of 1 kg from $9.89 to $10. There are 1000 g in a kilogram, so 750 g is about $\frac{3}{4}$ of a kilogram.

   Nelson buys about $\frac{3}{4}$ of a kilogram, so $\frac{3}{4}$ of $10 = $7.50 is our estimate.

   Step 4 Solve the problem. Use a unit fraction to convert 750 g to kilograms.

   $$\frac{750 \cancel{g}}{1} \cdot \frac{1 \text{ kg}}{1000 \cancel{g}} = 0.75 \text{ kg}$$

   Now multiply 0.75 kg times the cost per kilogram.

Now Try:

1. A ball of yarn weighs 140 g. Ellie knitted a sweater for her boyfriend that used 11 balls of yarn. How many kilograms did the finished sweater weigh?

   _____

$$\frac{0.75 \ \cancel{kg}}{1} \cdot \frac{\$9.89}{1 \ \cancel{kg}} = \$7.4175 \approx \$7.42$$

Step 5 State the answer. Nelson paid $7.42, rounded to the nearest cent.

Step 6 Check your work. The exact answer of $7.42 is close to our estimate of $7.50.

2. A 70-L drum is filled with oil which is to be packaged into 140-mL bottles. How many bottles can be filled.?

Step 1 Read the problem. The problem asks for the number of filled bottles.

Step 2 Work out a plan. The given amount is in liters, but the capacity of the bottles is in mL. Convert liters to milliliters, then divide by 140 mL (the capacity of the bottles).

Step 3 Estimate a reasonable answer. To estimate round 140 mL to 100 mL. Then 70 L = 70,000 mL, and $70,000 \div 100 = 700$ bottles.

Step 4 Solve the problem. On the metric conversion line, moving from L to mL is three places to the right, so move the decimal point in 70 L three places to the right. Then divide by 140.

70 L = 70,000 mL

$$\frac{70,000 \text{ mL}}{140 \text{ mL}} = 500 \text{ bottles}$$

Step 5 State the answer. 500 bottles will be filled.

Step 6 Check your work. The exact answer of 500 bottles is close to our estimate of 700 bottles.

3. Lucy purchased 1 m 60 cm of fabric at $6.25 per meter for a jacket and 1m 80 cm of fabric at $4.75 per m for a dress. How much did she spend in total?

Step 1 Read the problem. Lucy purchased two quantities of two different kinds of fabric. Find the total she spent on fabric.

Step 2 Work out a plan. The lengths involve two

2. If 1.8 kg of candy is to be divided equally among 9 children, how many grams will each child receive?

_____

3. A fish tank can hold up to 75.6 L of water. If there are 70,000 mL of water in the tank, how many more milliliters of water can the tank hold?

_____

Copyright © 2014 Pearson Education, Inc.

units, m and cm. Rewrite both lengths in meters, determine the price, and find the total.

Step 3 Estimate a reasonable answer. To estimate, 1 m 60 cm can be rounded to 2 m. Round 1 m 80 cm to 2 m. Also, $6.25 can be rounded to $6 and $4.75 can be rounded to $5. Then, $2 \times \$6 + 2 \times \$5 = \$12 + \$10 = \$22$, our estimate.

Step 4 Solve the problem. Rewrite the lengths in meters.

$$
\begin{array}{ll}
1 \text{ m} \rightarrow \quad 1.0 \text{ m} & 1 \text{ m} \rightarrow \quad 1.0 \text{ m} \\
\text{plus } 60 \text{ cm} \rightarrow \underline{+\ 0.6 \text{ m}} & \text{plus } 80 \text{ cm} \rightarrow \underline{+\ 0.8 \text{ m}} \\
\qquad\qquad 1.6 \text{ m} & \qquad\qquad 1.8 \text{ m}
\end{array}
$$

Jacket: $1.6 \times \$6.25 = \$10$
Dress: $1.8 \times \$4.75 = \underline{\$\ 8.55}$
$\qquad\qquad\qquad\qquad \$18.55$ total

Step 5 State the answer. The total spent is $18.55.

Step 6 Check your work. The exact answer of $18.55 is close to our estimate of $22.

Objective 1   Solve application problems involving metric measurements.

For extra help, see Examples 1–3 on pages 607–608 of your text and Section Lecture video for Section 8.4 and Exercise Solutions Clip 3, 9, 11, and 13.

Solve each application problem. Round money answers to the nearest cent.

1. Metal chain costs $5.26 per meter. Find the cost of 2 m 47 cm of the chain. Round your answer to the nearest cent.

    1. _____

2. Henry bowls with a 7-kg bowling ball, while Denise uses a ball that weighs 5 kg 750g. How much heavier is Henry's bowling ball than Denise's?

    2. _____

3. The label on a bottle of pills says that there are 3.5 mg of the medication in 5 pills. If a patient needs to take 8.4 mg of the medication, how many pills does he need to take?

    3. _____

Chapter 8 MEASUREMENT

8.5   Metric−U.S. Measurement Conversions and Temperature

Learning Objectives
1   Use unit fractions to convert between metric and U.S. measurement units.
2   Learn common temperatures on the Celsius scale.
3   Use formulas to convert between Celsius and Fahrenheit temperatures.

Key Terms

Use the vocabulary terms listed below to complete each statement in exercises 1−2.

Celsius   **Fahrenheit**

1.   The _____ scale is used to measure temperature in the metric system.

2.   The _____ scale is used to measure temperature in the U.S. customary system.

Guided Examples

Review these example for Objective 1:

1.   Convert from 16 in. to centimeters.

We're changing from a U.S length unit to a metric length unit. In the "U.S. to Metric Units" side of the table, you see that 1 inch ≈ 2.54 centimeters. Two unit fractions can be written using that information.

$$\frac{1 \text{ in.}}{2.54 \text{ cm}} \quad \text{or} \quad \frac{2.54 \text{ cm}}{1 \text{ in.}}$$

Multiply by the unit fraction that allows you to divide out inches.

$$16 \text{ in.} \cdot \frac{2.54 \text{ cm}}{1 \text{ in.}} = \frac{16 \text{ in.}}{1} \cdot \frac{2.54 \text{ cm}}{1 \text{ in.}} = 40.64 \text{ cm}$$

$$16 \text{ in.} \approx 40.64 \text{ cm}$$

Now Try:

1.   Convert from 12.9 m to feet.

_____

2. Convert using unit fractions. Round your answers to the nearest tenth.

    a. 22.5 pounds to kilograms

Look in the "U.S. to Metric Units" side of the table to see that 1 pound ≈ 0.45 kilograms . Use this information to write a unit fraction that allows you to divide out pounds.

$$\frac{22.5 \;\cancel{lb}}{1} \cdot \frac{0.45 \text{ kg}}{1 \;\cancel{lb}} = 10.125 \text{ kg}$$

22.5 lb ≈ 10.1 kg

    b. 46.8 L to quarts

Look in the "Metric to U.S. Units" side of the table to see that $1 \text{ L} \approx 1.06$ quarts . Write a unit fraction that allows you to divide out liters.

$$\frac{46.8 \;\cancel{L}}{1} \cdot \frac{1.06 \text{ qt}}{1 \;\cancel{L}} = 49.608 \text{ qt}$$

46.8 L ≈ 49.6 qt

Review these examples for Objective 2:

3. Choose the metric temperature that is most reasonable for each situation.

    a. hot water in a bathtub
        27°C   40°C   100°C

27°C is too cold and 100°C is too hot.
40°C is reasonable.

    b. fall day
        13°C   50°C   65°C

50°C and 65°C are too hot.
13°C is reasonable.

2. Convert using unit fractions. Round your answers to the nearest tenth.
    a. 9.68 kg to pounds

    _____

    b. 20 gallons to liters

    _____

Now Try:

3. Choose the metric temperature that is most reasonable for each situation.

    a. hot cocoa
    10°C   35°C   65°C

    _____

    b. ice cream
    −10°C   4°C   30°C

    _____

Review these examples for Objective 3:

4. Convert 77°F to Celsius.

   Use the formula and follow the order of operations.

   $$C = \frac{5(F-32)}{9}$$

   $$= \frac{5(77-32)}{9}$$

   $$= \frac{5(45)}{9}$$

   $$= \frac{5(\overset{5}{\cancel{45}})}{\underset{1}{\cancel{9}}}$$

   $$= 25$$

   Thus, 77°F = 25°C.

5. Convert 30°C to Fahrenheit.

   Use the formula and follow the order of operations.

   $$F = \frac{9 \cdot C}{5} + 32$$

   $$= \frac{9 \cdot 30}{5} + 32$$

   $$= \frac{9 \cdot \overset{6}{\cancel{30}}}{\underset{1}{\cancel{5}}} + 32$$

   $$= 54 + 32$$

   $$= 86$$

   Thus, 30°C = 86°F.

Now Try:

4. Convert 122°F to Celsius.

   _____

5. Convert 150°C to Fahrenheit.

   _____

Objective 1   Use unit fractions to convert between metric and U.S. measurement units.

For extra help, see Examples 1–2 on pages 611–612 of your text and Section Lecture video for Section 8.5 and Exercise Solutions Clip 7, 9, and 13.

Use the table in your textbook and unit fractions to make the following conversions. Round answers to the nearest tenth.

1. 291 mi to kilometers

   1. _____

2. 7 L to gallons

   2. _____

3.   26 oz to grams   3. _____

Objective 2   Learn common temperatures on the Celsius scale.

For extra help, see Example 3 on page 613 of your text and Section Lecture video for Section 8.5.

Choose the most reasonable temperature for each situation.

4.   hot coffee   4. _____
          35°C   60°C   100°C

5.   Normal body temperature   5. _____
          37°C   37°F

6.   Oven temperature   6. _____
          300°C   300°F

Objective 3   Use formulas to convert between Celsius and Fahrenheit temperatures.

For extra help, see Examples 4–5 on page 614 of your text and Section Lecture video for Section 8.5 and Exercise Solutions Clip 31.

Use the conversions formulas and the order of operations to convert Fahrenheit temperatures to Celsius and Celsius temperatures to Fahrenheit. Round your answers to the nearest degree, if necessary.

7.   62°F   7. _____

8.   10°C   8. _____

Solve the application problem. Round to the nearest degree, if necessary.

9.   A recipe for roast beef calls for an oven temperature   9. _____
          of 400°F. What is the temperature in degrees
          Celsius?

Chapter 9 GRAPHS AND GRAPHING

9.1   Problem Solving with Tables and Pictographs

Learning Objectives
1   Read and interpret data presented in a table.
2   Read and interpret data from a pictograph.

Key Terms

Use the vocabulary terms listed below to complete each statement in exercises 1−2.

    table   **pictograph**

1.   A graph that uses pictures or symbols to display information is called a

    _____.

2.   A display of facts in rows and columns is called a _____.

Guided Examples

The table below shows information about the performance of the eight U.S. airlines during the year 2011 (January−December).

PERFORMANCE DATA FOR SELECTED U.S. AIRLINES
January−December 2011

Airline	On-Time Performance	Luggage Handling*
American	78%	3.6
Atlantic Southeast	75%	5.5
Continental	77%	3.4
Delta	82%	2.7
Hawaiian	93%	2.6
JetBlue	73%	2.2
Southwest	81%	3.7
United	80%	3.7

*Luggage problems per 1000 passengers
Source: Department of Transportation Air Travel Consumer Report

Review these examples for Objective 1:

1.   Use the table above to answer these questions.

    a. What percent of United's flights were on time?

Look across the row labeled United. 80% of its flights were on time.

Now Try:

1.   Use the table above to answer these questions.

    a. What percent of Hawaiian's flights were on time?

_____

b. Which airline had the best luggage handling record?

Look down the column headed Luggage Handling. To find the best record, look for the lowest number, which is 2.2. Then look to the left to find the airline, which is JetBlue.

c. What was the average percent of on-time flights for the three airlines with the worst performance, to the nearest whole percent?

Look down the column headed On-Time Performance to find the three lowest numbers: 73%, 75%, and 77%. To find the average, add the values and divide by 3.

$$\frac{73+75+77}{3} = \frac{225}{3} = 75$$

The average on-time performance for the three worst airlines was 75%.

b. What airline(s) had the second worst luggage handling record?

_____

c. What was the average on-time performance for all eight airlines, to the nearest percent?

_____

The table below shows the maximum cab fares in five different cities in 2011. The "flag drop" change is made when the driver starts the meter. "Wait time" is the charge for having to wait in the middle of a ride.

MAXIMUM TAXICAB FARES ALLOWED IN SELECTED CITIES IN 2011

City	Flag Drop	Price per Mile	Wait Time (per Hour)
Chicago	$2.25	$1.80	$20
Denver	$2.50	$2.25	$24
Miami	$2.50	$2.40	$24
New York	$2.50	$2	$24
San Francisco	$3.50	$2.75	$27

Source: taxifarefinder.com

2. Use the table above to answer these questions.

a. What is the maximum fare for a 15-mile ride in Denver that includes having the cab wait 10 minutes?

The price per mile in Denver is $2.25, so the cost for 15 miles is 15($2.25) = $33.75. Then add the flag drop charge of $2.50. Finally, figure out the cost of the wait time. One way is to set up a proportion.

2. Use the table above to answer these questions.
a. What is the maximum fare for a 25-mile ride in Miami that includes having the cab wait 5 minutes?

_____

$$\text{Cost} \rightarrow \frac{\$24}{60 \text{ min}} = \frac{\$x}{10 \text{ min}} \leftarrow \text{Cost}$$
$$\text{Wait time} \rightarrow \quad \quad \quad \quad \leftarrow \text{Wait time}$$

$$60 \cdot x = 24 \cdot 10$$

$$\frac{60x}{60} = \frac{240}{60}$$

$$x = \$4$$

Total fare = $33.75 + $2.50 + $4 = $40.25

b. It is customary to give the cab driver a tip. Find the total cost of the cab ride in part (a) if the passenger added at 10% tip, rounded to the nearest quarter (nearest $0.25).

Use the percent equation to find the exact tip.

percent · whole = part

$$10\%(\$40.25) = n$$

$$0.10(\$40.25) = n$$

$$\$4.025 = n \quad \text{Round to } \$4 \text{ (nearest } \$0.25)$$

The total cost of the cab ride
is $40.25 + $4 tip = $44.25.

b. It is customary to give the cab driver a tip. Find the total cost of the cab ride in part (a) if the passenger added at 20% tip, rounded to the nearest quarter (nearest $0.25).

_____

This pictograph shows the approximate number of passenger arrivals and departures at selected U.S. airports in 2006.

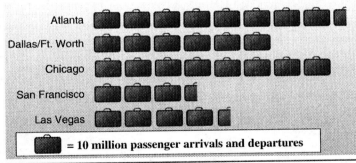

Source: Airports Council International—North America.

Review these examples for Objective 2:

3. Use the pictograph to answer these questions.

a. Approximately how many passenger arrivals and departures took place at the San Francisco airport?

The passenger arrivals and departures for San Francisco shows 3 whole symbols ($3 \cdot 10$ million = 30 million) plus half of a symbol ($\frac{1}{2}$ of 10 million is 5 million) for a total of 35 million.

Now Try:

3. Use the pictograph to answer these questions.

a. Approximately how many passenger arrivals and departures took place at the Las Vegas airport?

_____

Name: Date:
Instructor: Section:

b. What is the difference in the number of arrivals and departures at Dallas/Ft. Worth airport and Atlanta airport?

Dallas/Fort Worth shows 6 symbols and Atlanta shows $8\frac{1}{2}$ symbols. So Atlanta has $2\frac{1}{2}$ more symbols than Dallas/Fort Worth, and $2\frac{1}{2}\cdot 10$ million $= 25$ million. Thus, Atlanta has 25 million more passenger arrivals and departures than Dallas/Fort Worth.

b. What is the difference in the number of arrivals and departures at Chicago airport and San Francisco airport?

_____

Objective 1   Read and interpret data presented in a table.

For extra help, see Examples 1–2 on pages 634–635 of your text and Section Lecture video for Section 9.1 and Exercise Solutions Clip 13 and 19.

Use the table below for exercises 1–3. Use proportions to help solve exercise 3.

Weight of Exerciser	123 lbs	130 lbs	143 lbs
Calories burned in 30 minutes			
Cycling	168	177	195
Running	324	342	375
Jumping Rope	273	288	315
Walking	162	171	189

Source: Fitness magazine

1. How many calories are burned by a 130-pound adult in 30 minutes of cycling?

1. _____

2. Which activity will burn at least 300 calories when performed by a 123-pound adult?

2. _____

3. How many calories will a 123-pound adult burn in 60 minutes of walking and 15 minutes of running?

3. _____

Copyright © 2014 Pearson Education, Inc.

Name: Date:
Instructor: Section:

Objective 2   Read and interpret data from a pictograph.

For extra help, see Example 3 on page 636 of your text and Section Lecture video for Section 9.1 and Exercise Solutions Clip 27.

The pictograph below shows the population of five cities in 2010. Use the pictograph to answer exercises 4–6.

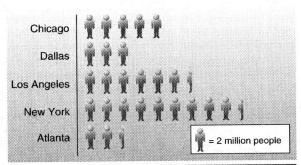

POPULATION OF U.S. METROPOLITAN AREAS

Source: U.S. Bureau of the Census, U.S. Department of Commerce.

4.   What is the approximate population of Dallas?   4. _____

5.   Approximately how much greater is the population   5. _____
     of Los Angeles than the population of Chicago?

6.   What is the approximate total population of all five   6. _____
     cities?

Chapter 9 GRAPHS AND GRAPHING

9.2   Reading and Constructing Circle Graphs

Learning Objectives
1   Read a circle graph.
2   Use a circle graph.
3   Use a protractor to draw a circle graph.

Key Terms

Use the vocabulary terms listed below to complete each statement in exercises 1−2.

circle graph   protractor

1.   A _____ shows how a total amount is divided into parts or sectors.

2.   A _____ is a device used to measure the number of degrees in angles or parts of a circle.

Guided Examples

The circle graph shows the number of students enrolled in certain majors at a college. The entire circle represents 11,600 students.

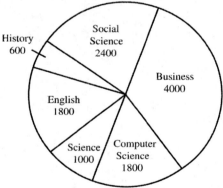

Review these examples for Objective 2:

1.   Find the ratio of social science majors to the total number of students. Write the ratio as a fraction in lowest terms.

The circle graph shows 2400 social science majors of the 11,600 total students. The ratio of social science students to the total number of students is shown below.

$$\frac{2400 \text{ students (social science)}}{11,600 \text{ students (total)}}$$

$$= \frac{2400 \text{ students}}{11,600 \text{ students}} = \frac{2400 \div 400}{11,600 \div 400} = \frac{6}{29}$$

Now Try:

1.   Find the ratio of history majors to the total number of students. Write the ratio as a fraction in lowest terms.

_____

2. Use the circle graph to find the ratio of history majors to business majors. Write the ratio as a fraction in lowest terms.

The circle graph shows 600 history majors and 4000 business majors. The ratio of history majors to business majors is shown below.

$$\frac{600 \text{ students (history)}}{4000 \text{ students (business)}}$$

$$= \frac{600 \text{ students}}{4000 \text{ students}} = \frac{600 \div 200}{4000 \div 200} = \frac{3}{20}$$

2. Use the circle graph to find the ratio of science majors to computer science majors. Write the ratio as a fraction in lowest terms.

_____

The circle graph shows the expenses involved in keeping a sales force on the road. Each expense item is expressed as a percent of the total sales force cost of $950,000.

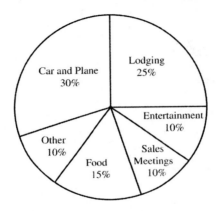

3. Use the circle graph above on expenses to find the amount spent on lodging.

Recall the percent equation.
    percent · whole = part
The total expenses is $950,000, so the whole is $950,000. The percent is 25%, or as a decimal, 0.25. Find the part.
    percent · whole = part
$$(0.25)(950,000) = n$$
$$237,500 = n$$
The amount spent on lodging was $237,500.

3. Use the circle graph above on expenses to find the amount spent on food.

_____

Review this example for Objective 3:

4. A family recorded its expenses for a year, with the following results: 40% for Housing, 20% for Food, 14% for Automobile, 8% for Clothing, 6% for Medical, 8% for Savings, and 4% for Other.

a. Find the number of degrees in a circle graph for each type of expenses.

Recall that a complete circle has 360°. Because "Housing" makes up 40% of the expenses, the number of degrees needed for the "Housing" sector of the circle graph is 40% of 360°.

Housing
$$(360°)(40\%) = (360°)(0.40) = 144°$$

Food
$$(360°)(20\%) = (360°)(0.20) = 72°$$

Automobile
$$(360°)(14\%) = (360°)(0.14) = 50.4°$$

Clothing
$$(360°)(8\%) = (360°)(0.08) = 28.8°$$

Medical
$$(360°)(6\%) = (360°)(0.06) = 21.6°$$

Savings
$$(360°)(8\%) = (360°)(0.08) = 28.8°$$

Other
$$(360°)(4\%) = (360°)(0.04) = 14.4°$$

b. Draw a circle graph showing this information.

Use a protractor to make the circle graph.

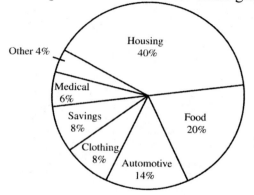

Now Try:

4. A book publisher had 30% of its sales in mysteries, 15% in biographies, 10% in cookbooks, 25% in romance novels, 15% in science, and the rest in business books.

a. Find the number of degrees in a circle graph for each type of book.

mysteries _____

biographies _____

cookbooks _____

romance_____

science_____

business _____

b. Draw a circle graph showing this information.

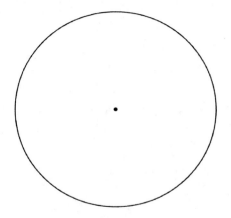

Name: Date:

Instructor: Section:

Objective 1   Read a circle graph.

For extra help, see page 641 of your text and Section Lecture video for Section 9.2.

The circle graph shows the cost of remodeling a kitchen. Use the graph to answer exercises 1–3.

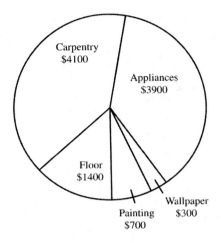

1. Find the total cost of remodeling the kitchen.   1. _____

2. What is the largest single expense in remodeling the   2. _____
    kitchen?

3. How much less does the wallpaper cost than   3. _____
    painting?

Objective 2   Use a circle graph.

For extra help, see Examples 1–3 on pages 641–642 of your text and Section Lecture video for Section 9.2 and Exercise Solutions Clip 21, 22, and 23.

The circle graph shows the number of students enrolled in certain majors at a college. Use the graph to answer exercises 4–5. The entire circle represents 11,600 students.

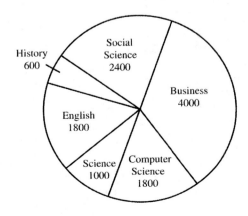

4.   Find the ratio of the number of business majors to the total number of students.

4. _____

5.   Find the ratio of the number of science majors to the number of English majors.

5. _____

The circle graph shows the expenses involved in keeping a sales force on the road. Each expense item is expressed as a percent of the total sales force cost of $950,000. Find the number of dollars of expense for the category.

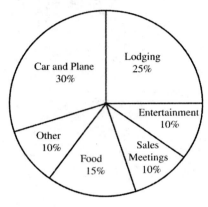

6.   Car and plane

6. _____

Objective 3   Use a protractor to draw a circle graph.

For extra help, see Example 4 on pages 643–644 of your text and Section Lecture video for Section 9.2 and Exercise Solutions Clip 31.

Use the given information to draw a circle graph.

Jensen Manufacturing Company has its annual sales divided into five categories as follows. The total sales for a year is $400,000.

Item	Annual Sales
Parts	$20,000
Hand tools	80,000
Bench tools	100,000
Brass fittings	140,000
Cabinet hardware	60,000

7. Find the percent of the total sales for each item.

7. parts _____

hand tools _____

bench tools _____

brass fittings _____

hardware _____

8. Find the number of degrees in a circle graph for each item.

8. parts _____

hand tools _____

bench tools _____

brass fittings _____

hardware _____

9. Make a circle graph showing this information.

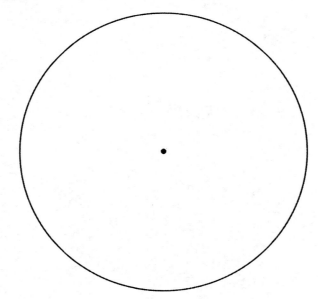

Chapter 9 GRAPHS AND GRAPHING

9.3   Bar Graphs and Line Graphs

Learning Objectives
1   Read and understand a bar graph.
2   Read and understand a double-bar graph.
3   Read and understand a line graph.
4   Read and understand a comparison line graph.

Key Terms

Use the vocabulary terms listed below to complete each statement in exercises 1–4.

bar graph   double-bar graph   line graph   comparison line graph

1.   A _____ uses dots connected by a line to show trends.

2.   A _____ compares two sets of data by showing two sets of bars.

3.   A _____ uses bars of various heights or lengths to show quantity or frequency.

4.   A _____ shows how two sets of data relate to each other by showing a line graph for each item.

Guided Examples

The bar graph shows the enrollment for the fall semester at a small college from 2002 to 2006.

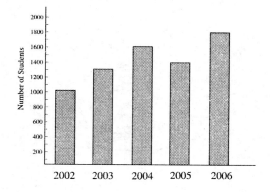

Review this example for Objective 1:

1.   What was the fall enrollment for 2003?

    The bar for 2003 rises to 1300. So the enrollment in 2003 was 1300 students.

Now Try:

1.   What was the fall enrollment for 2006?

    _____

Name: Date:
Instructor: Section:

The double-bar graph shows the enrollment by gender in each class at a small college.

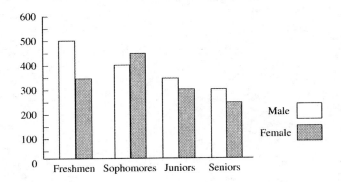

Review these examples for Objective 2:

2. Use the double-bar graph to find the following.

a. The number of male sophomores enrolled

There are two bars for the sophomore class. The color code to the right tells you that white bars represent male enrollments. So the white bar on the left for Sophomores represents the number of male sophomores. It rises to 400.
So the sophomore enrollment of males is 400.

b. The number of female juniors enrolled

The gray column for juniors rises to 300.
So the junior enrollment of females is 300.

Now Try:

2. Use the double-bar above to find the following.

a. The number of female seniors enrolled

_____

b. The number of male freshmen enrolled

_____

The line graph gives the value of one share of stock of Microchip Computer Corporation on the first trading day of the month for six consecutive months.

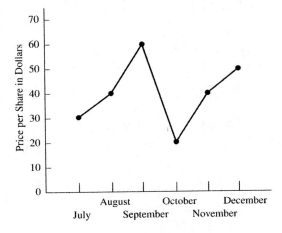

Review these examples for Objective 3:

3. Use the line graph on the previous page to find the following.

a. In which month was the value of the stock the lowest?

The lowest point on the graph is the dot directly over October, so the lowest value of the stock occurred in October.

b. Find the value of one share of stock on the first trading day in August.

Use a ruler or straightedge to line up the August dot with the numbers along the left edge of the graph. The August dot is directly across from 40. So in August, the price per share was $40.

Now Try:

3. Use the line graph on the previous page to find the following.
a. Find the value of the stock in September.

_____

b. In which month was the value of the stock $50?

_____

The comparison line graph shows annual sales for two different stores from 2002 to 2006.

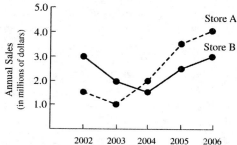

Review these examples for Objective 4:

4. Use the comparison line graph to find the following.

a. The number of annual sales for store A in 2004

Find the dot on the dotted line above 2004. It rises to $2.0. Multiply 2.0 by $1,000,000 because the label on the left side of the graph says in millions of dollars.
So, in 2004 the annual sales for store A was $2,000,000.

b. The number of annual sales for store B in 2006

The solid line on the graph shows 3.0 times $1,000,000 or $3,000,000 was the annual sales for store B in 2006.

Now Try:

4. Use the comparison line graph to find the following.

a. The number of annual sales for store B in 2004

_____

b. The number of annual sales for store A in 2006

_____

c. Find the amount of decrease and the percent of decrease in Store B's annual sales from 2003 to 2004. Round to the nearest whole percent, if necessary.

Use subtraction to find the amount of decrease.

$$\begin{array}{ccc} \text{Store B} & & \text{Store B} & & \text{amount} \\ \text{sales in 2003} & - & \text{sales in 2004} & = & \text{of decrease} \end{array}$$

$$2{,}000{,}000 \quad - \quad 1{,}500{,}000 \quad = \quad 500{,}000$$

Now use the percent equation to find the percent of decrease.

percent of original sales = amount of decrease

$$p \cdot 2{,}000{,}000 = 500{,}000$$

$$\frac{p \cdot 2{,}000{,}000}{2{,}000{,}000} = \frac{500{,}000}{2{,}000{,}000}$$

$$p = 0.25$$

$$p = 0.25 = 25\%$$

The amount of decrease is $500,000.
The percent of decrease is 25%.

c. The amount of decrease and the percent of decrease in Store B's annual sales from 2002 to 2003. Round to the nearest whole percent, if necessary.

_____

_____

Objective 1   Read and understand a bar graph.

For extra help, see Example 1 on page 650 of your text and Section Lecture video for Section 9.3.

The bar graph shows the enrollment for the fall semester at a small college from 2002 to 2006. Use this graph for problems 1−3.

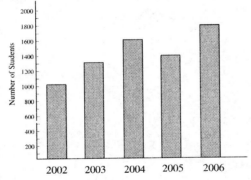

1.   What was the fall enrollment for 2004?

1. _____

2.   What year had the greatest enrollment?

2. _____

3.   By how many students did the enrollment increase from 2005 to 2006?

3. _____

Name: Date:
Instructor: Section:

Objective 2   Read and understand a double-bar graph.

For extra help, see Example 2 on page 651 of your text and Section Lecture video for Section 9.3.

The double-bar graph shows the enrollment by gender in each class at a small college. Use the double-bar graph for Problems 4–6.

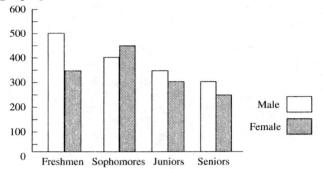

4.   How many female freshmen are enrolled?   4. _____

5.   Which class has a greater female enrollment than   5. _____
      male enrollment?

6.   Find the total number of juniors enrolled.   6. _____

Objective 3   Read and understand a line graph.

For extra help, see Example 3 on page 651 of your text and Section Lecture video for Section 9.3 and Exercise Solutions Clip 25 and 27.

The line graph gives the value of one share of stock of Microchip Computer Corporation on the first trading day of the month for six consecutive months. Use the line graph for Problems 7–9.

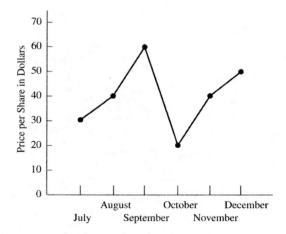

7.   In which month was the value of the stock highest?   7. _____

8. Find the value of one share on the first trading day October.

8. _____

9. By how much did the value of one share increase from July to September?

9. _____

Objective 4   Read and understand a comparison line graph.

For extra help, see Example 4 on page 652 of your text and Section Lecture video for Section 9.3 and Exercise Solutions Clip 33.

The comparison line graph shows annual sales for two different stores from 2002 to 2006. Use the graph to solve Problems 10–12.

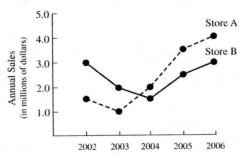

10. Find the annual sales for store A in 2005.

10. _____

11. Find the annual sales for store B in 2003.

11. _____

12. Find the amount of decrease and the percent of decrease in Store A's annual sales from 2002 to 2003. Round to the nearest whole percent, if necessary.

12. _____

Chapter 9 GRAPHS AND GRAPHING

9.4   The Rectangular Coordinate System

Learning Objectives
1   Plot a point, given the coordinates, and find the coordinates, given a point.
2   Identify the four quadrants and determine which points lie within each one.

Key Terms

Use the vocabulary terms listed below to complete each statement in exercises 1−8.

paired data	**horizontal axis**	**vertical axis**
ordered pair	**x-axis**	**y-axis**
coordinate system **origin**	**coordinates**	**quadrants**

1.   An _____ is the "address" of a point in a coordinate system.

2.   In a rectangular coordinate system, the axis that goes "left and right" is called the
      _____ or the _____.

3.   In a rectangular coordinate system, the axis that goes "up and down" is called the
      _____ or the _____.

4.   When each number in a set of data is matched with another number by some rule
      of association, we call it _____.

5.   Together, the x-axis and the y-axis form a rectangular _____.

6.   The x-axis and the y-axis divide the coordinate system into four regions called
      _____.

7.   The axis lines in a coordinate system intersect at the _____.

8.   _____ are the numbers in the ordered pair that specify
      the location of a point on a rectangular coordinate system.

Name: Date:

Instructor: Section:

Guided Examples

Review these examples for Objective 1:

1. Use the grid below to plot each point.

    a. (2, 5)

    Start at (0, 0). Move to the right along the
    horizontal axis until you reach 2. Then move up
    5 units so that you are aligned with 5 on the
    vertical axis. Make a dot and label it (2, 5).

    b. (−5, 4)

    Start at (0, 0). Move to the left along the
    horizontal axis until you reach −5. Then move up
    4 units so that you are aligned with 4 on the
    vertical axis. Make a dot and label it (−5, 4).

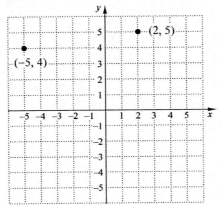

2. Use the grid below to plot each point.

    a. (1, −2)

    Move right 1 unit. Move down 2 units. Make a
    dot and label it (1, −2).

    b. (−4, 0)

    Move left 4 units. Make a dot and label it (−4, 0).

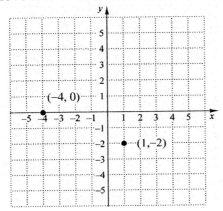

Now Try:

1. Use the grid to plot each point.

    a. (1, 3)

    b. (4, −1)

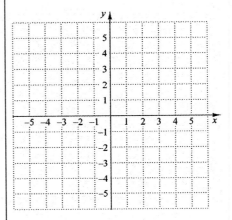

2. Use the grid to plot each point.

    a. (−2, 0)

    b. (0, 5)

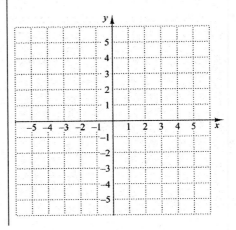

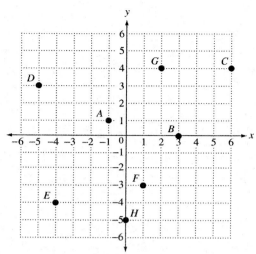

3. Give the coordinates of each point.

    a. *A*

    To reach point *A* from the origin, move 1 unit to the left; then move up 1 unit. The coordinates are (–1, 1).

    b. *B*

    To reach point *B* from the origin, move 3 units to the left; do not move up or down. The coordinates are (3, 0).

    c. *C*

    To reach point *C* from the origin, move 6 units to the right; then move up 4 units. The coordinates are (6, 4).

    d. *D*

    To reach point *D* from the origin, move 5 units to the left; then move up 3 units. The coordinates are (–5, 3).

3. Give the coordinates of each point.

    a. *E*

    _____

    b. *F*

    _____

    c. *G*

    _____

    d. *H*

    _____

Review these examples for Objective 2:

4. Identify the quadrant in which each point is located.

    a. (5, 11)

    For (5, 11), the pattern is (+, +), so the point is in Quadrant I.

    b. (3, –2)

    For (3, –2), the pattern is (+, –), so the point is in Quadrant IV.

Now Try:

4. Identify the quadrant in which each point is located.

    a. (4, 9)

    _____

    b. (–6, 3)

    _____

                Copyright © 2014 Pearson Education, Inc.

c. (–8, 0)

The point corresponding to (–8, 0) is on
the *x*-axis, so it isn't in any quadrant.

c. (0, 9)

_____

Objective 1   **Plot a point, given the coordinates, and find the coordinates, given a point.**

For extra help, see Examples 1–3 on pages 658–660 of your text and Section Lecture video for Section 9.4 and Exercise Solutions Clip 5 and 7.

1. *Plot each point on the rectangular coordinate system below. Label each point with its coordinates.*

a.   (–3, 4)

b.   (2, –1)

c.   (0, –5)

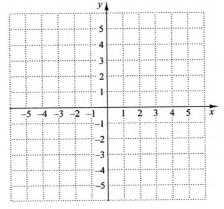

2. *Give the coordinates of each point.*

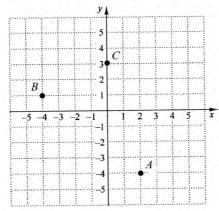

A _____

B _____

C _____

Objective 2   **Identify the four quadrants and determine which points lie within each one.**

For extra help, see Example 4 on page 660 of your text and Section Lecture video for Section 9.4 and Exercise Solutions Clip 9 and 11.

Identify the quadrant in which each point is located.

3.   (3, –9)

4.   (–2, –2)

5.   (–3, 0)

3. _____

4. _____

5. _____

Chapter 9 GRAPHS AND GRAPHING

9.5   Introduction to Graphing Linear Equations

Learning Objectives
1   Graph linear equations in two variables.
2   Identify the slope of a line as positive or negative.

Key Terms

Use the vocabulary terms listed below to complete each statement in exercises 1–2.

> **graph a linear equation**   **slope**

1.   To _____, find at least three ordered pairs that satisfy the equation, and then, plot the points and connect them with a straight line.

2.   As you move from left to right, if a line tilts downward, it has a negative _____.

Guided Examples

Review these examples for Objective 1:

1.   Graph $x + y = 7$ by finding three solutions and plotting the ordered pairs. Then use the graph to find a fourth solution of the equation.

Set up a table to organize the information.

x	y	$x + y = 7$	(x, y)
2	5	$2 + 5 = 7$	$(2, 5)$
3	4	$3 + 4 = 7$	$(3, 4)$
4	3	$4 + 3 = 7$	$(4, 3)$

Plot the ordered pairs and draw a line through the points, extending it in both directions.

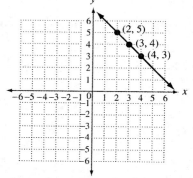

Find another point. It appears that a point with coordinates $(5, 2)$ is a solution.

Now Try:

1.   Graph $x - y = 0$ by finding three solutions and plotting the ordered pairs. Then use the graph to find a fourth solution of the equation.

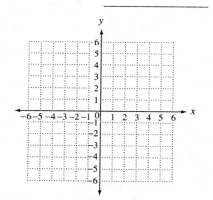

To check that (5, 2) is a solution, substitute 5 for x and 2 for y in the original equation.

$$x + y = 7$$

$$5 + 2 = 7$$

$$7 = 7$$

The equation balances, so (5, 2) is another solution of $x + y = 7$.

2. Graph $y = 3x$ by finding three solutions and plotting the ordered pairs. Then use the graph to find a fourth solution of the equation.

Set up a table to organize the information.

x	y	$y = 3x$	(x, y)
-1	-3	$-3 = 3(-1)$	$(-1, -3)$
0	0	$0 = 3(0)$	$(0, 0)$
1	3	$3 = 3(1)$	$(1, 3)$

Plot the ordered pairs and draw a line through the points, extending it in both directions.

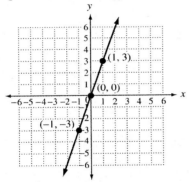

Find another point. It appears that a point with coordinates (2, 6) is a solution.

To check that (2, 6) is a solution, substitute 2 for x and 6 for y in the original equation.

$$y = 3x$$

$$6 = 3(2)$$

$$6 = 6$$

The equation balances, so (2, 6) is another solution of $y = 3x$.

2. Graph $y = 4x - 1$ by finding three solutions and plotting the ordered pairs. Then use the graph to find a fourth solution of the equation.

_____

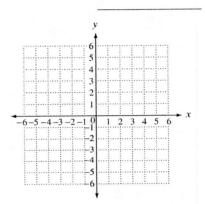

3. Graph $y = -\dfrac{1}{4}x$ by finding three solutions and plotting the ordered pairs. Then use the graph to find a fourth solution of the equation.

Set up a table to organize the information.

x	y	$y = -\dfrac{1}{4}x$	(x, y)
-4	1	$1 = -\dfrac{1}{4}(-4)$	$(-4, 1)$
0	0	$0 = -\dfrac{1}{4}(0)$	$(0, 0)$
4	-1	$-1 = -\dfrac{1}{4}(4)$	$(4, -1)$

Plot the ordered pairs and draw a line through the points, extending it in both directions.

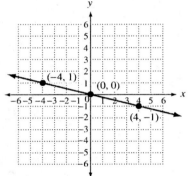

Find another point. It appears that a point with coordinates $\left(-2, \dfrac{1}{2}\right)$ is a solution.

To check that $\left(-2, \dfrac{1}{2}\right)$ is a solution, substitute -2 for x and $\dfrac{1}{2}$ for y in the original equation.

$$y = -\dfrac{1}{4}x$$

$$\dfrac{1}{2} = -\dfrac{1}{4}(-2)$$

$$\dfrac{1}{2} = \dfrac{1}{2}$$

The equation balances, so $\left(-2, \dfrac{1}{2}\right)$ is another

solution of $y = -\dfrac{1}{4}x$.

3. Graph $y = \dfrac{1}{5}x$ by finding three solutions and plotting the ordered pairs. Then use the graph to find a fourth solution of the equation.

_____

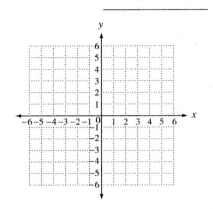

4. Graph $y = -3x + 2$ by finding three solutions and plotting the ordered pairs. Then use the graph to find a fourth solution of the equation.

Set up a table to organize the information.

x	y	$y = -3x + 2$	(x, y)
0	2	$2 = -3(0) + 2$	$(0, 2)$
1	-1	$-1 = -3(1) + 2$	$(1, -1)$
2	-4	$-4 = -3(2) + 2$	$(2, -4)$

Plot the ordered pairs and draw a line through the points, extending it in both directions.

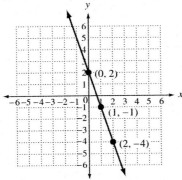

Find another point. It appears that a point with coordinates $(-1, 5)$ is a solution.

To check that $(-1, 5)$ is a solution, substitute -1 for x and 5 for y in the original equation.

$$y = -3x + 2$$
$$5 = -3(-1) + 2$$
$$5 = 5$$

The equation balances, so $(-1, 5)$ is another solution of $y = -3x + 2$.

4. Graph $y = x + 2$ by finding three solutions and plotting the ordered pairs. Then use the graph to find a fourth solution of the equation.

_____

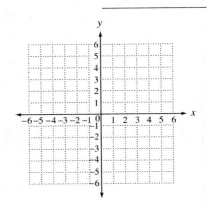

Review this example for Objective 2:

5. Look back at the graph of $y = -3x + 2$ from Example 4. Then complete these sentences.
The graph of $y = -3x + 2$ has a _____ slope.
As the value of x increases, the value of y ____.

The graph of $y = -3x + 2$ has a <u>negative</u> slope (because it tilts downward).
As the value of x increases, the value of y <u>decreases</u> (does the opposite).

Now Try:

5. Look back at the graph of $y = x + 2$ from Example 4.
Then complete these sentences.
The graph of $y = x + 2$ has a
_____ slope. As the value of x
increases, the value of y ____.

_____

_____

Name: Date:

Instructor: Section:

Objective 1   Graph linear equations in two variables.

For extra help, see Examples 1–4 on pages 664–668 of your text and Section Lecture video for Section 9.5 and Exercise Solutions Clip 5, 9, 13, and 17.

Graph each equation. Make your own table using the listed values of x.

1. $y = x - 4$
      Use 0, 2, and 4 as the values of x.

1.

2. $x + y = -3$
      Use –4, –3, and –2 as the values of x.

2.

Objective 2   Identify the slope of a line as positive or negative.

For extra help, see Example 5 on page 670 of your text and Section Lecture video for Section 9.5.

Look back at the graphs in exercises 1–2. Then complete exercises 3–4.

3.   The graph of $y = x - 4$ has a _____ slope.   3. _____

4.   The graph of $x + y = -3$ has a _____ slope.   4. _____

Chapter 10 EXPONENTS AND POLYNOMIALS

10.1   The Product Rule and Power Rules for Exponents

Learning Objectives	
1	Review the use of exponents.
2	Use the product rule for exponents.
3	Use the rule $\left(a^m\right)^n = a^{mn}$.
4	Use the rule $\left(ab\right)^m = a^m b^m$.
5	Use the rule $\left(\dfrac{a}{b}\right)^m = \dfrac{a^m}{b^m}$.

Key Terms

Use the vocabulary terms listed below to complete each statement in exercises 1–3.

exponential expression   **base**   **power**

1. 2^5 is read "2 to the fifth _____".

2. A number written with an exponent is called a(n)
    _____.

3. The _____ is the number being multiplied repeatedly.

Guided Examples

Review these examples for Objective 1:

1. Write $5 \cdot 5 \cdot 5$ in exponential form and evaluate.

    Since 5 occurs as a factor three times, the base is 5 and the exponent is 3.
    $$5 \cdot 5 \cdot 5 = 5^3 = 125$$

2. Name the base and exponent of each expression. Then evaluate.

    a. 3^4

    Base: 3
    Exponent: 4
    Value: $3^4 = 3 \cdot 3 \cdot 3 \cdot 3 = 81$

    b. -3^4

    Base: 3
    Exponent: 4
    Value: $-3^4 = -1 \cdot (3 \cdot 3 \cdot 3 \cdot 3) = -81$

Now Try:

1. Write $4 \cdot 4 \cdot 4 \cdot 4 \cdot 4$ in exponential form and evaluate.

    _____

2. Name the base and exponent of each expression. Then evaluate.

    a. 2^6

    _____

    b. -2^6

    _____

c. $(-3)^4$

Base: –3
Exponent: 4
Value: $(-3)^4 = (-3)(-3)(-3)(-3) = 81$

c. $(-2)^6$

_____

Review these examples for Objective 2:

3. Use the product rule for exponents to simplify, if possible.

a. $8^4 \cdot 8^5$

$8^4 \cdot 8^5 = 8^{4+5}$

$= 8^9$

b. $(-5)^8 (-5)^3$

$(-5)^8 (-5)^3 = (-5)^{8+3}$

$= (-5)^{11}$

c. $x^3 \cdot x$

$x^3 \cdot x = x^3 \cdot x^1$

$= x^{3+1}$

$= x^4$

d. $m^7 m^8 m^9$

$m^7 m^8 m^9 = m^{7+8+9}$

$= m^{24}$

e. $5^2 \cdot 4^3$

$5^2 \cdot 4^3 = 25 \cdot 64$

$= 1600$

f. $5^2 + 5^3$

$5^2 + 5^3 = 25 + 125$

$= 150$

Now Try:

3. Use the product rule for exponents to simplify, if possible.

a. $9^6 \cdot 9^7$

_____

b. $(-6)^2 (-6)^4$

_____

c. $x^5 \cdot x$

_____

d. $m^{11} m^9 m^7$

_____

e. $5^3 \cdot 2^2$

_____

f. $3^4 + 3^3$

_____

4. Multiply $5x^4$ and $6x^9$.

$$5x^4 \cdot 6x^9 = (5 \cdot 6) \cdot \left(x^4 \cdot x^9\right)$$
$$= 30x^{4+9}$$
$$= 30x^{13}$$

4. Multiply $6x^5$ and $3x^6$.

_____

Review these examples for Objective 3:

5. Use power rule (a) for exponents to simplify.

a. $\left(3^4\right)^5$

$$\left(3^4\right)^5 = 3^{4 \cdot 5}$$
$$= 3^{20}$$

b. $\left(5^6\right)^3$

$$\left(5^6\right)^3 = 5^{6 \cdot 3}$$
$$= 5^{18}$$

c. $\left(x^4\right)^4$

$$\left(x^4\right)^4 = x^{4 \cdot 4}$$
$$= x^{16}$$

d. $\left(m^3\right)^4$

$$\left(m^3\right)^4 = m^{3 \cdot 4}$$
$$= m^{12}$$

Now Try:

5. Use power rule (a) for exponents to simplify.

a. $\left(4^5\right)^3$

_____

b. $\left(7^2\right)^4$

_____

c. $\left(x^7\right)^7$

_____

d. $\left(n^5\right)^6$

_____

Review these examples for Objective 4:

6. Use power rule (b) for exponents to simplify.

a. $(5xy)^3$

$$(5xy)^3 = 5^3 x^3 y^3$$
$$= 125x^3 y^3$$

b. $6(rs)^4$

$$6(rs)^4 = 6\left(r^4 s^4\right)$$
$$= 6r^4 s^4$$

Now Try:

6. Use power rule (b) for exponents to simplify.

a. $(4ab)^3$

_____

b. $7(pq)^3$

_____

c. $4\left(3m^4 n^5\right)^3$

$$4\left(3m^4 n^5\right)^3 = 4\left[3^3 \left(m^4\right)^3 \left(n^5\right)^3\right]$$
$$= 4\left[3^3 m^{12} n^{15}\right]$$
$$= 108 m^{12} n^{15}$$

c. $8\left(5x^4 y^3\right)^2$

_____

Review these examples for Objective 5:

7. Use power rule (c) for exponents to simplify.

a. $\left(\dfrac{3}{5}\right)^4$

$$\left(\dfrac{3}{5}\right)^4 = \dfrac{3^4}{5^4} = \dfrac{81}{625}$$

b. $\left(\dfrac{x}{y}\right)^5$ (where $y \neq 0$)

$$\left(\dfrac{x}{y}\right)^5 = \dfrac{x^5}{y^5}$$

Now Try:

7. Use power rule (c) for exponents to simplify.

a. $\left(\dfrac{5}{6}\right)^3$

_____

b. $\left(\dfrac{p}{q}\right)^6$

_____

Objective 1   Review the use of exponents.

For extra help, see Examples 1–2 on page 696 of your text and Section Lecture video for Section 10.1.

Write the expression in exponential form and evaluate, if possible.

1. $\left(\dfrac{1}{3}\right)\left(\dfrac{1}{3}\right)\left(\dfrac{1}{3}\right)\left(\dfrac{1}{3}\right)\left(\dfrac{1}{3}\right)$

1. _____

Evaluate each exponential expression. Name the base and the exponent.

2. $(-4)^4$

2. _____

base_____

exponent_____

3. -3^8

3. _____

base_____

exponent_____

Name: Date:
Instructor: Section:

Objective 2   Use the product rule for exponents.

For extra help, see Examples 3–4 on pages 697–698 of your text and Section Lecture video for Section 10.1 and Exercise Solutions Clip 27, 29, and 35.

Use the product rule to simplify each expression, if possible. Write each answer in exponential form.

4. $7^4 \cdot 7^3$ 4. _____

5. $(-2c^7)(-4c^8)$ 5. _____

6. $(3k^7)(-8k^2)(-2k^9)$ 6. _____

Objective 3   Use the rule $\left(a^m\right)^n = a^{mn}$.

For extra help, see Example 5 on page 698 of your text and Section Lecture video for Section 10.1 and Exercise Solutions Clip 41.

Simplify each expression. Write all answers in exponential form.

7. $\left(7^3\right)^4$ 7. _____

8. $-\left(v^4\right)^9$ 8. _____

9. $\left[2^3\right]^7$ 9. _____

Objective 4   Use the rule $\left(ab\right)^m = a^m b^m$.

For extra help, see Example 6 on page 699 of your text and Section Lecture video for Section 10.1 and Exercise Solutions Clip 49 and 57.

Simplify each expression.

10. $\left(5r^3t^2\right)^4$ 10. _____

11. $\left(3a^4b\right)^3$ 11. _____

12. $\left(-2w^3z^7\right)^4$

12. _____

Objective 5   Use the rule $\left(\dfrac{a}{b}\right)^m = \dfrac{a^m}{b^m}$ **.**

For extra help, see Example 7 on page 700 of your text and Section Lecture video for Section 10.1 and Exercise Solutions Clip 51.

Simplify each expression.

13. $\left(-\dfrac{2x}{5}\right)^3$

13. _____

14. $\left(\dfrac{xy}{z^2}\right)^4$

14. _____

15. $\left(\dfrac{-2a}{b^2}\right)^7$

15. _____

Chapter 10 EXPONENTS AND POLYNOMIALS

10.2 Integer Exponents and the Quotient Rule

Learning Objectives	
1	Use 0 as an exponent.
2	Use negative numbers as exponents.
3	Use the quotient rule for exponents.
4	Use the product rule with negative exponents.

Key Terms

Use the vocabulary terms listed below to complete each statement in exercises 1−3.

exponent   **base**   **product rule for exponents**

power rule for exponents

1. The statement "If m and n are any integers, then $\left(a^m\right)^n = a^{mn}$ " is an example of the _____.

2. In the expression a^m, a is the _____ and m is the _____.

3. The statement "If m and n are any integers, then $a^m \cdot a^n = a^{m+n}$ is an example of the _____.

Guided Examples

Review these examples for Objective 1:

1. Evaluate.

    a. $75^0 = 1$

    b. $(-75)^0 = 1$

    c. $-75^0 = -(1)$ or -1

    d. $x^0 = 1$ $(x \neq 0)$

Now Try:

1. Evaluate.

    a. 88^0

    b. $(-88)^0$

    c. -88^0

    d. a^0 $(a \neq 0)$

e. $9x^0 = 9(1)$, or 9 $(x \neq 0)$

f. $(9x)^0 = 1$ $(x \neq 0)$

e. $88a^0$ $(a \neq 0)$

_____

f. $(88a)^0$ $(a \neq 0)$

_____

Review these examples for Objective 2:

2. Simplify by writing each expression with positive exponents. Then evaluate the expression, if possible.

a. 5^{-2}

$5^{-2} = \dfrac{1}{5^2}$,   or $\dfrac{1}{25}$

b. 4^{-3}

$4^{-3} = \dfrac{1}{4^3}$,   or $\dfrac{1}{64}$

c. $5^{-1} - 3^{-1}$

$$5^{-1} - 3^{-1} = \frac{1}{5} - \frac{1}{3}$$
$$= \frac{3}{15} - \frac{5}{15}$$
$$= -\frac{2}{15}$$

d. q^{-3}

$q^{-3} = \dfrac{1}{q^3}$

Now Try:

2. Simplify by writing each expression with positive exponents. Then evaluate the expression, if possible.

a. 8^{-2}

_____

b. 3^{-3}

_____

c. $4^{-1} - 8^{-1}$

_____

d. p^{-4}

_____

Review these examples for Objective 3:

3. Simplify. Assume that all variables represent nonzero real numbers.

a. $\dfrac{4^9}{4^6} = 4^{9-6} = 4^3 = 64$

b. $\dfrac{5^3}{5^7} = 5^{3-7} = 5^{-4} = \dfrac{1}{5^4} = \dfrac{1}{625}$

Now Try:

3. Simplify. Assume that all variables represent nonzero real numbers.

a. $\dfrac{3^{18}}{3^{16}}$

_____

b. $\dfrac{2^6}{2^{10}}$

_____

c. $\dfrac{3^{-4}}{3^{-9}} = 3^{-4-(-9)} = 3^5 = 243$

c. $\dfrac{8^{-4}}{8^{-6}}$

_____

d. $\dfrac{p^6}{p^{-4}} = p^{6-(-4)} = p^{10}$

d. $\dfrac{z^6}{z^{-4}}$

_____

Review these examples for Objective 4:

4. Simplify each expression. Assume that all variables represent nonzero real numbers. Write answers with positive exponents.

a. $10^6\left(10^{-5}\right)$

$10^6\left(10^{-5}\right) = 10^{6+(-5)} = 10^1 \text{ or } 10$

b. $\left(10^{-7}\right)\left(10^{-4}\right)$

$\left(10^{-7}\right)\left(10^{-4}\right) = 10^{-7+(-4)} = 10^{-11} = \dfrac{1}{10^{11}}$

c. $y^{-3} \cdot y^6 \cdot y^{-9}$

$$y^{-3} \cdot y^6 \cdot y^{-9} = y^{-3+6} \cdot y^{-9}$$
$$= y^3 \cdot y^{-9}$$
$$= y^{3+(-9)}$$
$$= y^{-6}$$
$$= \dfrac{1}{y^6}$$

Now Try:

4. Simplify each expression. Assume that all variables represent nonzero real numbers. Write answers with positive exponents.

a. $10^7\left(10^{-6}\right)$

_____

b. $\left(10^{-11}\right)\left(10^{-3}\right)$

_____

c. $y^{-6} \cdot y^5 \cdot y^{-7}$

_____

Objective 1   Use 0 as an exponent.

For extra help, see Example 1 on page 703 of your text and Section Lecture video for Section 10.2.

Evaluate each expression.

1. -12^0

1. _____

2. $-15^0 - (-15)^0$

2. _____

3. $\dfrac{0^8}{8^0}$ 3. _____

Objective 2   Use negative numbers as exponents.

For extra help, see Example 2 on page 704 of your text and Section Lecture video for Section 10.2 and Exercise Solutions Clip 21 and 23.

Evaluate or simplify each expression, and write it using only positive exponents. Assume that all variables represent nonzero real numbers.

4. 2^{-4} 4. _____

5. n^{-9} 5. _____

6. 3^{-1} 6. _____

Objective 3   Use the quotient rule for exponents.

For extra help, see Example 3 on page 705 of your text and Section Lecture video for Section 10.2 and Exercise Solutions Clip 29, 33, and 41.

Use the quotient rule to simplify each expression, and write it using only positive exponents. Assume that all variables represent nonzero real numbers.

7. $\dfrac{13^4}{13^{-3}}$ 7. _____

8. $\dfrac{m^{-3}}{m^{-8}}$ 8. _____

9. $\dfrac{z^{-17}}{z^{19}}$ 9. _____

Name: _____ Date: _____

Instructor: _____ Section: _____

Objective 4 Use the product rule with negative exponents.

For extra help, see Example 4 on page 706 of your text and Section Lecture video for Section 10.2 and Exercise Solutions Clip 57, 63, and 65.

Simplify each expression, and write it using only positive exponents. Assume that all variables represent nonzero real numbers.

10. $3^{-9}\left(3^4\right)$

10. _____

11. $p^{-7} \cdot p^4 \cdot p^{15}$

11. _____

12. $x^{-3} \cdot x^2 \cdot x^{-11}$

12. _____

Chapter 10 EXPONENTS AND POLYNOMIALS

10.3   An Application of Exponents: Scientific Notation

Learning Objectives
1	Express numbers in scientific notation.
2	Convert numbers in scientific notation to numbers without exponents.
3	Use scientific notation in calculations.
4	Solve application problems using scientific notation.

Key Terms

Use the vocabulary terms listed below to complete each statement in exercises 1–3.

   scientific notation   **quotient rule**   **power rule**

1.   _____ is used to write very large or very small numbers using exponents.

2.   The statement "If m and n are any integers and $b \neq 0$, then $\left(\dfrac{a}{b}\right)^m = \dfrac{a^m}{b^m}$" is an

   example of the _____.

3.   The statement "If m and n are any integers and $b \neq 0$, then $\dfrac{a^m}{a^n} = a^{m-n}$" is an

   example of the _____.

Guided Examples

Review these examples for Objective 1:

1.   Write each number in scientific notation.

   a. 54,000,000

   Move the decimal point to follow the 5. Count the number of places the decimal point was moved: 7 places.

   $54,000,000 = 5.4 \times 10^7$

   b. 84,300,000,000

   Move the decimal point 10 places to the left.

   $84,300,000,000 = 8.43 \times 10^{10}$

   c. 7.406

   $7.406 = 7.406 \times 10^0$

Now Try:

1.   Write each number in scientific notation.

   a. 680,000,000

   _____

   b. 47,710,000,000

   _____

   c. 9.991

   _____

d. 0.00573

The first nonzero digit is 5. Count the places.
Move the decimal point 3 places to the right.

$$0.00573 = 5.73 \times 10^{-3}$$

e. −0.0000617

Move the decimal point 5 places to the right.

$$-0.0000617 = -6.17 \times 10^{-5}$$

d. 0.0463

_____

e. −0.000848

_____

Review these examples for Objective 2:
2. Write each number without exponents.

a. 7.4×10^4

Move the decimal point 4 places to the right, and add three zeros.

$$7.4 \times 10^4 = 74,000$$

b. 3.57×10^6

Move the decimal point 6 places to the right, and add four zeros.

$$3.57 \times 10^6 = 3,570,000$$

c. -8.98×10^{-3}

Move the decimal point 3 places to the left.

$$-8.98 \times 10^{-3} = -0.00898$$

Now Try:
2. Write each number without exponents.

a. 8.35×10^5

_____

b. 2.796×10^7

_____

c. -1.64×10^{-4}

_____

Review these examples for Objective 3:
3. Perform each calculation. Write answers in scientific notation and also without exponents.

a. $\left(8 \times 10^4\right)\left(7 \times 10^3\right)$

$$\left(8 \times 10^4\right)\left(7 \times 10^3\right) = (8 \times 7)\left(10^4 \times 10^3\right)$$
$$= 56 \times 10^7$$
$$= \left(5.6 \times 10^1\right) \times 10^7$$
$$= 5.6 \times 10^8$$
$$= 560,000,000$$

Now Try:
3. Perform each calculation. Write answers in scientific notation and also without exponents.

a. $\left(9 \times 10^5\right)\left(3 \times 10^2\right)$

_____

b. $\dfrac{6\times10^{-4}}{3\times10^{2}}$

$$\dfrac{6\times10^{-4}}{3\times10^{2}}=\dfrac{6}{3}\times\dfrac{10^{-4}}{10^{2}}$$

$$=2\times10^{-6}$$

$$=0.000002$$

b. $\dfrac{39\times10^{-3}}{13\times10^{5}}$

_____

Review these examples for Objective 4:

4. In 2005, there were about 102 people per square mile living in Wisconsin. The state has an area of 5.43×10^{4} square miles. What was the population of Wisconsin?

   First write 102 in scientific notation.
   $$102=1.02\times10^{2}$$
   Now multiply the number of people in one square mile times the number of square miles.
   $$\left(1.02\times10^{2}\right)\left(5.43\times10^{4}\right)$$
   $$=\left(1.02\times5.43\right)\left(10^{2}\times10^{4}\right)$$
   $$=5.5386\times10^{6}$$
   Thus, there are 5.5386×10^{6} people in Wisconsin.

5. In 2007, the Gross Domestic Product of the U.S. was 1.38×10^{13} dollars. That year the U.S. population was about 3.02×10^{8} people. What was the per capita GDP, the Gross Domestic Product per person? Round to the nearest hundred.

   Divide the GDP by the population.
   $$\dfrac{1.38\times10^{13}}{3.02\times10^{8}}=\dfrac{1.38}{3.02}\times\dfrac{10^{13}}{10^{8}}$$
   $$=0.45695\times10^{5}$$
   $$=4.5695\times10^{4}\text{ or }45,695$$
   Rounded to the nearest hundred, the GDP is $45,700 per person.

Now Try:

4. There are about 6×10^{23} atoms in a mole of atoms. How many atoms are there in 8.1×10^{-5} mole?

   _____

5. Earth has a mass of 6×10^{24} kilograms and a volume of 1.1×10^{21} cubic meters. What is Earth's density in kilograms per cubic meter? Round to the nearest hundredth.

   _____

Name: Date:

Instructor: Section:

Objective 1   Express numbers in scientific notation.

For extra help, see Example 1 on pages 711–712 of your text and Section Lecture video for Section 10.3 and Exercise Solutions Clip 17 and 21.

Write each number in scientific notation.

1.   23,651   1. _____

2.   −429,600,000,000   2. _____

3.   −0.0002208   3. _____

Objective 2   Convert numbers in scientific notation to numbers without exponents.

For extra help, see Example 2 on page 712 of your text and Section Lecture video for Section 10.3 and Exercise Solutions Clip 23 and 29.

Write each number without exponents.

4. -2.45×10^6 4. _____

5. 6.4×10^{-3} 5. _____

6. -4.02×10^0 6. _____

Objective 3   Use scientific notation in calculations.

For extra help, see Example 3 on page 712 of your text and Section Lecture video for Section 10.3 and Exercise Solutions Clip 33, 37, and 39.

Perform the indicated operations, and write the answers in scientific notation.

7. $\left(2.3 \times 10^4\right) \times \left(1.1 \times 10^{-2}\right)$ 7. _____

8. $\dfrac{9.39 \times 10^1}{3 \times 10^3}$ 8. _____

Objective 4   Solve application problems using scientific notation.

For extra help, see Examples 4–5 on pages 713–714 of your text and Section Lecture video for Section 10.3.

Work each problem. Give answers both in scientific notation and without exponents.

9. Americans eat 6.1 million pretzels in a day. How   9. _____
    many pretzels to Americans eat in 1 year?

10. An electronic oscillator is producing a signal with a   10. _____
    frequency of 1.7×10^8 cycles per second. This
    frequency is to be tripled. What will the new
    frequency be?

11. The moon has a mass of 7.35×10^{22} kilograms and a   11. _____
    volume of 2.2×10^{10} cubic meters. What is the
    moon's density in kilograms per cubic meter? Round
    to the nearest hundredth.

Chapter 10 EXPONENTS AND POLYNOMIALS

10.4   Adding and Subtracting Polynomials

Learning Objectives
1   Review combining like terms.
2   Use the vocabulary for polynomials.
3   Evaluate polynomials.
4   Add polynomials.
5   Subtract polynomials.

Key Terms

Use the vocabulary terms listed below to complete each statement in exercises 1–7.

> polynomial   descending powers   degree of a term
>
> degree of a polynomial   monomial
>
> binomial   trinomial

1. The _____ is the sum of the exponents on the variables in that term.

2. A polynomial in x is written in _____ if the exponents on x decrease from left to right.

3. A polynomial with exactly three terms is called a _____.

4. A _____ is a term, or the sum of terms, with whole number exponents.

5. A polynomial with exactly one term is called a _____.

6. The _____ is the highest degree of any term of the polynomial.

7. A _____ is a polynomial with exactly two terms.

Guided Examples

Review these examples for Objective 1:

1. Simplify each expression by combining like terms.

    a. $-7x^2 + 10x^2$

    $-7x^2 + 10x^2 = (-7 + 10)x^2$

    $= 3x^2$

Now Try:

1. Simplify each expression by combining like terms.

    a. $-6x^4 + 11x^4$

    _____

b. $4x^5 - 15x^5 + x^5$

$4x^5 - 15x^5 + x^5 = (4 - 15 + 1)x^5$

$\qquad = -10x^5$

c. $19m^3 + 6m + 5m^3$

$19m^3 + 6m + 5m^3 = (19 + 5)m^3 + 6m$

$\qquad = 24m^3 + 6m$

d. $5rs^3 + rs^3 - 4rs^3$

$5rs^3 + rs^3 - 4rs^3 = (5 + 1 - 4)rs^3$

$\qquad = 2rs^3$

b. $9x^7 - 18x^7 + x^7$

_____

c. $22m^2 + 15m^3 + 7m^2$

_____

d. $6p^2q - 3p^2q + 2p^2q$

_____

Review these examples for Objective 2:

2. Simplify each polynomial if possible. Then give the degree and tell whether the polynomial is a monomial, a binomial, a trinomial, or none of these.

a. $5x^4 + 7x$

We cannot simplify further. This is a binomial of degree 4.

b. $9x - 7x + 3x$

$9x - 7x + 3x = 5x$
The degree is 1. The simplified polynomial is a monomial.

Now Try:

2. Simplify each polynomial if possible. Then give the degree and tell whether the polynomial is a monomial, a binomial, a trinomial, or none of these.

a. $8x^3 + 4x^2 + 6$

_____

b. $x^5 + 3x^5$

_____

Review this example for Objective 3:

3. Find the value of $4x^3 + 6x^2 - 5x - 5$ when $x = -3$ and $x = 2$.

For $x = -3$, replace x with –3.
$4x^3 + 6x^2 - 5x - 5$

$= 4(-3)^3 + 6(-3)^2 - 5(-3) - 5$

$= 4(-27) + 6(9) - 5(-3) - 5$

$= -108 + 54 + 15 - 5$

$= -44$

Now Try:

3. Find the value of
$5x^4 + 3x^2 - 9x - 7$ when $x = 4$
and $x = -4$.

_____

_____

Now replace x with 2.

$4x^3 + 6x^2 - 5x - 5$

$= 4(2)^3 + 6(2)^2 - 5(2) - 5$

$= 4(8) + 6(4) - 5(2) - 5$

$= 32 + 24 - 10 - 5$

$= 41$

Review these examples for Objective 4:

4. Add vertically.

a. $5x^4 - 7x^3 + 9$ and $-3x^4 + 8x^3 - 7$

Write like terms in columns.

$$5x^4 - 7x^3 + 9$$
$$-3x^4 + 8x^3 - 7$$

Now add, column by column.

$$\begin{array}{ccc} 5x^4 & -7x^3 & 9 \\ -3x^4 & 8x^3 & -7 \\ \hline 2x^4 & x^3 & 2 \end{array}$$

Add the three sums together to obtain the answer.

$$2x^4 + x^3 + 2$$

b. $3x^3 + 7x + 5$ and $x^4 - 6x$

Write like terms and add column by column.

$$\begin{array}{cccc} & 3x^3 & +7x & +5 \\ x^4 & & -6x & \\ \hline x^4 & +3x^3 & +x & +5 \end{array}$$

5. Find each sum by adding horizontally.

a. Add $5x^4 - 7x^3 + 9$ and $-3x^4 + 8x^3 - 7$

$\left(5x^4 - 7x^3 + 9\right) + \left(-3x^4 + 8x^3 - 7\right)$

$= 5x^4 - 3x^4 - 7x^3 + 8x^3 + 9 - 7$

$= 2x^4 + x^3 + 2$

Now Try:

4. Add vertically.

a. $8x^3 - 9x^2 + x$ and $-3x^3 + 4x^2 + 3x$

_____

b. $9x^4 + 3x - 6$ and $7x^2 + 4x$

_____

5. Find each sum by adding horizontally.

a. Add $15x^3 - 5x + 3$ and $-11x^3 + 6x + 9$

_____

b. $\left(5x^4 - 7x^2 + 6x\right) + \left(-3x^3 + 4x^2 - 7\right)$

$\left(5x^4 - 7x^2 + 6x\right) + \left(-3x^3 + 4x^2 - 7\right)$

$= 5x^4 - 3x^3 - 7x^2 + 4x^2 + 6x - 7$

$= 5x^4 - 3x^3 - 3x^2 + 6x - 7$

b. $\left(8x^2 - 6x + 4\right) + \left(7x^3 - 8x - 5\right)$

_____

Review these examples for Objective 5:

6. Perform each subtraction.

a. $(8x - 3) - (4x - 7)$

Use the definition of subtraction and the distributive property.

$(8x - 3) - (4x - 7) = (8x - 3) + [-(4x - 7)]$
$= (8x - 3) + [-1(4x - 7)]$
$= (8x - 3) + (-4x + 7)$
$= 4x + 4$

b. Subtract $8x^3 - 5x^2 + 8$ from $9x^3 + 6x^2 - 7$.

$\left(9x^3 + 6x^2 - 7\right) - \left(8x^3 - 5x^2 + 8\right)$
$= \left(9x^3 + 6x^2 - 7\right) + \left(-8x^3 + 5x^2 - 8\right)$
$= x^3 + 11x^2 - 15$

Now Try:

6. Perform each subtraction.

a. $(7x - 6) - (5x - 8)$

_____

b. $\left(7x^3 - 3x - 5\right) - \left(18x^3 + 4x - 6\right)$

_____

Objective 1   Review combining like terms.

For extra help, see Example 1 on page 717 of your text and Section Lecture video for Section 10.4 and Exercise Solutions Clip 23.

In each polynomial, combine like terms whenever possible. Write the result with descending powers.

1. $7z^3 - 4z^3 + 5z^3 - 11z^3$

1. _____

2. $-1.3z^7 + 0.4z^7 + 2.6z^8$

2. _____

3. $6c^3 - 9c^2 - 2c^2 + 14 + 3c^2 - 6c - 8 + 2c^3$

3. _____

Objective 2   Use the vocabulary for polynomials.

For extra help, see Example 2 on page 718 of your text and Section Lecture video for Section 10.4 and Exercise Solutions Clip 29 and 31

For each polynomial, first simplify, if possible, and write the resulting polynomial in descending powers of the variable. Then give the degree of this polynomial, and tell whether it is a monomial, *a* binomial, *a* trinomial, *or* none of these.

4. $3n^8 - n^2 - 2n^8$

4.   _____

degree: _____

type: _____

5. $-d^2 + 3.2d^3 - 5.7d^8 - 1.1d^5$

5.   _____

degree: _____

type: _____

6. $-6c^4 - 6c^2 + 9c^4 - 4c^2 + 5c^5$

6.   _____

degree: _____

type: _____

Objective 3   Evaluate polynomials.

For extra help, see Example 3 on page 718 of your text and Section Lecture video for Section 10.4 and Exercise Solutions Clip 41.

Find the value of each polynomial (a) *when x = –2 and* (b) *when x = 3.*

7. $3x^3 + 4x - 19$

7. a._____

b._____

8. $-4x^3 + 10x^2 - 1$

8. a._____

b._____

9. $x^4 - 3x^2 - 8x + 9$

9. a._____

b._____

Objective 4   Add polynomials.

For extra help, see Examples 4–5 on page 719 of your text and Section Lecture video for Section 10.4.

Add.

10. $9m^3 + 4m^2 - 2m + 3$

          $\underline{-4m^3 - 6m^2 - 2m + 1}$

10. _____

11. $\left(x^2 + 6x - 8\right) + \left(3x^2 - 10\right)$

11. _____

12. $\left(3r^3 + 5r^2 - 6\right) + \left(2r^2 - 5r + 4\right)$

12. _____

Objective 5   Subtract polynomials.

For extra help, see Example 6 on page 720 of your text and Section Lecture video for Section 10.4 and Exercise Solutions Clip 53.

Subtract.

13. $\left(-8w^3 + 11w^2 - 12\right) - \left(-10w^2 + 3\right)$

13. _____

14. $\left(8b^4 - 4b^3 + 7\right) - \left(2b^2 + b + 9\right)$

14. _____

15. $\left(9x^3 + 7x^2 - 6x + 3\right) - \left(6x^3 - 6x + 1\right)$

15. _____

Chapter 10 EXPONENTS AND POLYNOMIALS

10.5   Multiplying Polynomials: An Introduction

Learning Objectives
1   Multiply a monomial and a polynomial.
2   Multiply two polynomials.

Key Terms

Use the vocabulary terms listed below to complete each statement in exercises 1–5.

    monomial   **binomial**   **trinomial**

    polynomial   **distributive property**

1.   A _____ is a polynomial with exactly one term.

2. $a(b+c) = ab + ac$ is the statement of the _____.

3.   A _____ is a polynomial with exactly three terms.

4.   An algebraic expression made up of a term, or the sum of terms, with whole number exponents is called a _____.

5.   A polynomial with exactly two terms is a _____.

Guided Examples

Review these examples for Objective 1:
1.   Find each product.

    a. $5x^2(7x+3)$

Use the distributive property.
$$5x^2(7x+3) = 5x^2(7x) + 5x^2(3)$$
$$= 35x^3 + 15x^2$$

b. $-9n^4(6n^4 + 5n^3 + 7n - 1)$

Use the distributive property.
$$-9n^4(6n^4 + 5n^3 + 7n - 1)$$
$$= -9n^4(6n^4) - 9n^4(5n^3) - 9n^4(7n) - 9n^4(-1)$$
$$= -54n^8 - 45n^7 - 63n^5 + 9n^4$$

Now Try:
1.   Find each product.

    a. $8x^3(4x+8)$

    _____

    b. $-7m^5(5m^3 - 6m^2 + 4m - 1)$

    _____

Review these examples for Objective 2:

2. Find each product.

   a. Multiply $(x+7)(x-5)$.

Multiply each term of the second polynomial by each terms of the first. Then combine like terms.

$$(x+7)(x-5) = x(x) + x(-5) + 7(x) + 7(-5)$$
$$= x^2 + (-5x) + 7x + (-35)$$
$$= x^2 + 2x - 35$$

   b. Multiply $(x^2+6)(5x^3 - 4x^2 + 3x)$.

Multiply each term of the second polynomial by each term of the first.

$$(x^2+6)(5x^3 - 4x^2 + 3x)$$
$$= x^2(5x^3) + x^2(-4x^2) + x^2(3x)$$
$$\quad + 6(5x^3) + 6(-4x^2) + 6(3x)$$
$$= 5x^5 - 4x^4 + 3x^3 + 30x^3 - 24x^2 + 18x$$
$$= 5x^5 - 4x^4 + 33x^3 - 24x^2 + 18x$$

3. Multiply $(2x^3 + 7x^2 + 5x - 1)(4x + 6)$ using the vertical method.

Write the polynomials vertically.

$$2x^3 + 7x^2 + 5x - 1$$
$$\underline{\qquad\qquad\quad 4x + 6}$$

Begin by multiplying each term in the top row by 6.

$$2x^3 \;\; + 7x^2 \;\; + 5x \;\; - 1$$
$$\underline{\qquad\qquad\qquad 4x \;\; + 6}$$
$$12x^3 + 42x^2 + 30x - 6$$

Now multiply each term in the top row by $4x$. Then add like terms.

$$2x^3 \;\; + 7x^2 \;\; + 5x \;\; - 1$$
$$\underline{\qquad\qquad\qquad 4x \;\; + 6}$$
$$12x^3 \;\; + 42x^2 + 30x - 6$$
$$\underline{8x^4 + 28x^3 + 20x^2 \;\; - 4x}$$
$$8x^4 + 40x^3 + 62x^2 + 26x - 6$$

The product is $8x^4 + 40x^3 + 62x^2 + 26x - 6$.

Now Try:

2. Find each product.

   a. Multiply $(x+9)(x-6)$.

_____

   b. Multiply
   $(x^3+9)(4x^4 - 2x^2 + x)$.

_____

3. Multiply
$(4x^3 - 3x^2 + 6x + 5)(7x - 3)$
using the vertical method.

_____

Name: Date:

Instructor: Section:

Objective 1   Multiply a monomial and a polynomial.

For extra help, see Example 1 on page 725 of your text and Section Lecture video for Section 10.5 and Exercise Solutions Clip 13.

Find each product.

1. $7z(5z^3 + 2)$ 1. _____

2. $2m(3 + 7m^2 + 3m^3)$ 2. _____

3. $-3y^2(2y^3 + 3y^2 - 4y + 11)$ 3. _____

Objective 2   Multiply two polynomials.

For extra help, see Examples 2–3 on page 726 of your text and Section Lecture video for Section 10.5 and Exercise Solutions Clip 15, 17, 19, and 27.

Find each product.

4. $(x + 3)(x^2 - 3x + 9)$ 4. _____

5. $(2m^2 + 1)(3m^3 + 2m^2 - 4m)$ 5. _____

6. $(x - 5)(x + 4)$ 6. _____

Chapter R WHOLE NUMBERS REVIEW

R.1   Adding Whole Numbers

Learning Objectives	
1	Add two or three single-digit numbers.
2	Add more than two numbers.
3	Add when regrouping is not required.
4	Add with regrouping.
5	Use addition to solve application problems.
6	Check the sum in addition.

Key Terms

Use the vocabulary terms listed below to complete each statement in exercises 1–6.

> **addition   addends   sum (total)   commutative property of addition**
>
> **associative property of addition   regrouping**

1.   By the_____, changing the order of the addends in an addition problem does not change the sum.

2.   In addition, the numbers being added are called the _____.

3.   The process of finding the total is called _____.

4.   By the_____, changing the grouping of the addends in an addition problem does not change the sum.

5.   If the sum of the digits in any column is greater than 9, use the process called _____.

6.   The answer to an addition problem is called the _____.

Guided Examples

Review these examples for Objective 1:

1.   Add, and then change the order of numbers to write another addition problem.

    a. $4 + 5$

$4 + 5 = 9$ and $5 + 4 = 9$

    b. $3 + 7$

$3 + 7 = 10$ and $7 + 3 = 10$

Now Try:

1.   Add, and then change the order of numbers to write another addition problem.

    a. $7 + 2$

    _____

    b. $5 + 9$

    _____

Name: Date:
Instructor: Section:

c. $5 + 5$	**c.** $3 + 3$
$5 + 5 = 10$	_____

Review this example for Objective 2:	**Now Try:**
2. Add 7, 2, 5, 3, and 8.	2. Add 9, 7, 2, 5, and 6.

$$
\begin{array}{l}
7 \\
2 \quad 7+2=9 \\
5 \qquad\quad 9+5=14 \\
3 \qquad\qquad\quad 14+3=17 \\
\underline{+8} \qquad\qquad\qquad\quad 17+8=25 \\
25
\end{array}
$$

_____

Review this example for Objective 3:	**Now Try:**
3. Add 302, 23, 151, and 21.	3. Add 610, 42, 105, and 31.

First line up the numbers in columns, with the ones column at the right.

$$
\begin{array}{r}
302 \\
23 \\
151 \\
\underline{+\ 21}
\end{array}
$$

_____

Start at the right and add the ones digits. Add the tens digits next, and finally the hundreds digits.

$$
\begin{array}{r}
302 \\
23 \\
151 \\
\underline{+\ 21} \\
497
\end{array}
$$

Review these examples for Objective 4:	**Now Try:**
4. Add 38 and 27.	4. Add 46 and 19.

Add ones.

$$
\begin{array}{r}
38 \\
\underline{+\ 27}
\end{array}
$$

_____

8 ones and 7 ones = 15 ones

Regroup 15 ones as 1 ten and 5 ones. Write 5 ones in the ones column and write 1 ten in the tens column.

$$
\begin{array}{r}
1 \\
38 \\
\underline{+\ 27} \\
5
\end{array}
$$

Add the digits in the tens column, including the

regrouped 1.

$$
\begin{array}{r}
\overset{1}{38} \\
+\ 27 \\
\hline
65
\end{array}
\quad 1 \text{ ten} + 3 \text{ tens} + 2 \text{ tens} = 6 \text{ tens}
$$

5. Add 3197 + 420 + 638 + 67.

Step 1 Add the digits in the ones column.

$$
\begin{array}{r}
\overset{2}{} \\
3197 \\
420 \\
638 \\
+\ 67 \\
\hline
2
\end{array}
$$

Write 2 tens in the tens column.

Write 2 ones in the ones column.

Notice that 22 ones are regrouped as 2 tens and 2 ones.

Step 2 Add the digits in the tens column, including the regrouped 2.

$$
\begin{array}{r}
\overset{2\,2}{} \\
3197 \\
420 \\
638 \\
+\ 67 \\
\hline
22
\end{array}
$$

Write 2 hundreds in the hundreds column.

Write 2 tens in the tens column.

Step 3 Add the hundreds column, including the regrouped 2.

$$
\begin{array}{r}
\overset{1\,2\,2}{} \\
3197 \\
420 \\
638 \\
+\ 67 \\
\hline
322
\end{array}
$$

Write 1 thousand in the thousands column.

Write 3 hundreds in the hundreds column.

Step 4 Add the thousands column, including the regrouped 1.

$$
\begin{array}{r}
\overset{1\,2\,2}{} \\
3197 \\
420 \\
638 \\
+\ 67 \\
\hline
4322
\end{array}
$$

Sum of the thousands column is 4.

Finally, 3197 + 420 + 638 + 67 = 4322.

5. Add 6835 + 97 + 246 + 4001.

_____

Using the map below, find the shortest distance between the following cities. All distances are in miles.

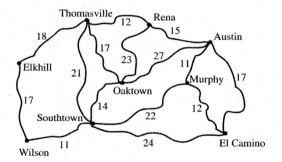

Review these examples for Objective 5:

6. Oaktown and El Camino

    One way from Oaktown to El Camino is through Austin. Add the mileage numbers along this route.

27	Oaktown to Austin
+ 17	Austin to El Camino
44	

    44 miles from Oaktown to El Camino going through Austin.

    Another way from Oaktown to El Camino is through Southtown. Add the mileage numbers along this route.

14	Oaktown to Southtown
+ 24	Southtown to El Camino
38	

    38 miles from Oaktown to El Camino going through Southtown.

    The shortest route from Oaktown to El Camino is 38 miles through Southtown.

7. Murphy and Thomasville

    Add the mileage from Murphy to Southtown to Thomasville.

11	Murphy to Austin
15	Austin to Rena
+ 12	Rena to Thomasville
38	

    38 miles from Murphy to Thomasville.

Now Try:

6. Southtown and Austin

    _____

7. El Camino and Thomasville

    _____

Review this example for Objective 6:

8. Check the sum. If the sum is incorrect, find the correct sum.

$$\begin{array}{r} 73 \\ 9815 \\ 390 \\ +\ 7002 \\ \hline 16,270 \end{array}$$

To check, add up.

$$\begin{array}{r} \underline{17,280} \\ 73 \\ 9815 \\ 390 \\ +\ 7002 \\ \hline 16,270 \end{array}$$

Re-add to find the sum is 17,280.

Now Try:

8. Check the sum. If the sum is incorrect, find the correct sum.

$$\begin{array}{r} 723 \\ 681 \\ 29 \\ 412 \\ +\ 103 \\ \hline 1947 \end{array}$$

_____

Objective 1   Add two or three single-digit numbers.

For extra help, see Example 1 on page 740 of your text and Section Lecture video for Section R.1.

Add.

1. $3 + 9$

2. $8 + 7$

1. _____

2. _____

Objective 2   Add more than two numbers.

For extra help, see Example 2 on page 740 of your text and Section Lecture video for Section R.1.

Add.

3. $\begin{array}{r} 3 \\ 4 \\ 9 \\ 2 \\ 7 \\ +\ 6 \end{array}$

3. _____

4.   6
      4
      9
      8
      4
  + 3

4. _____

Objective 3   Add when regrouping is not required.

For extra help, see Example 3 on page 741 of your text and Section Lecture video for Section R.1.

Add.

5.   421
     14
    532
  + 20

5. _____

6.   158
    340
  + 401

6. _____

Objective 4   Add with regrouping.

For extra help, see Examples 4–5 on pages 741–742 of your text and Section Lecture video for Section R.1.

Add.

7.   563
  + 478

7. _____

8.   7439
  + 8376

8. _____

9.   7033
    809
  2532
  +   41

9. _____

Name: Date:
Instructor: Section:

Objective 5   Use addition to solve application problems.

For extra help, see Examples 6–7 on page 743 of your text and Section Lecture video for Section R.1.

Using the map below, find the shortest distance between the following cities. All distances are in miles.

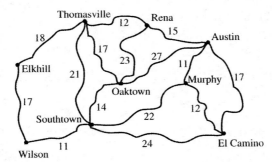

10.   Murphy and Oaktown   10. _____

11.   Wilson and Austin   11. _____

Objective 6   Check the sum in addition.

For extra help, see Example 8 on page 744 of your text and Section Lecture video for Section R.1.

Check the following additions. If an answer is incorrect, give the correct answer.

12.   67   12. _____
          48
       + 83
       ─────
        198

13.   3028   13. _____
         335
        2914
         688
      + 1647
      ──────
        8482

Chapter R WHOLE NUMBERS REVIEW

R.2   Subtracting Whole Numbers

Learning Objectives
1   Change addition problems to subtraction problems or the reverse.
2   Identify the minuend, subtrahend, and difference.
3   Subtract when no regrouping is needed.
4   Use addition to check subtraction answers.
5   Subtract with regrouping.
6   Use subtraction to solve application problems.

Key Terms

Use the vocabulary terms listed below to complete each statement in exercises 1–4.

       minuend   **subtrahend**   **difference**   **regrouping**

1.   The number from which another number is being subtracted is called the

    _____.

2.   In order to subtract 29 from 76, use a process called _____ from the tens place.

3.   The _____ is the number being subtracted.

4.   The answer to a subtraction problem is called the _____.

Guided Examples

Review these examples for Objective 1:

1.   Change each addition problem to a subtraction problem.

    a. $4 + 2 = 6$

    Two subtraction problems are possible:
    $6 - 2 = 4$ or $6 - 4 = 2$
    b. $5 + 9 = 14$

    $14 - 5 = 9$ or $14 - 9 = 5$

2.   Change each subtraction problem to an addition problem.

    a. $9 - 4 = 5$

    Two addition problems are possible:
    $9 = 4 + 5$ or $9 = 5 + 4$
    b. $390 - 150 = 240$

    $390 = 150 + 240$ or $390 = 240 + 150$

Now Try:

1.   Change each addition problem to a subtraction problem.

    a. $8 + 5 = 13$

    _____

    b. $9 + 3 = 12$

    _____

2.   Change each subtraction problem to an addition problem.

    a. $9 - 7 = 2$

    _____

    b. $320 - 190 = 130$

    _____

Review this example for Objective 2:

3. Identify the minuend, subtrahend, and difference in the following subtraction problem.

    94 − 13 = 81

    The number 94 is the minuend.
    The number 13 is the subtrahend.
    The number 81 is the difference.

Now Try:

3. Identify the minuend, subtrahend, and difference in the following subtraction problem.

    87 − 12 = 75

    Minuend _____

    Subtrahend _____

    Difference _____

Review these examples for Objective 3:

4. Subtract.

    a.   63
          − 31

    3 ones − 1 one = 2 ones
    6 tens − 3 tens = 3 tens
           63
         − 31
         ____
           32

    b.   463
          − 143

    3 ones − 3 ones = 0 ones
    6 tens − 4 tens = 2 tens
    4 hundreds − 1 hundred = 3 hundreds
           463
         − 143
         ____
           320

    c.   9546
          − 305

    6 ones − 5 ones = 1 ones
    4 tens − 0 tens = 4 tens
    5 hundreds − 3 hundreds = 2 hundreds
    9 thousands − 0 thousand = 9 thousand
          9546
         − 305
         ____
          9241

Now Try:

4. Subtract.

    a.   85
          − 42
          ____

    b.   538
          − 427
          ____

    c.   6574
          − 364
          ____

Review these examples for Objective 4:

5. Use addition to check each answer. If the answer is incorrect, find the correct answer.

a.
$$
\begin{array}{r}
58 \\
-\ 37 \\
\hline 21
\end{array}
$$

Rewrite as an addition problem.
$$
\begin{array}{r}
37 \\
+\ 21 \\
\hline 58
\end{array}
$$

Because $37 + 21 = 58$, the subtraction was done correctly.

b. $83 - 32 = 52$

Rewrite as an addition problem.
$$32 + 52 = 83$$
But $32 + 52 = 84$, not 83, so the subtraction was done incorrectly. Reworking the subtraction,
$$83 - 32 = 51.$$

c.
$$
\begin{array}{r}
785 \\
-\ 162 \\
\hline 623
\end{array}
$$

Since $162 + 623 = 785$, the answer checks.

Now Try:

5. Use addition to check each answer. If the answer is incorrect, find the correct answer.

a.
$$
\begin{array}{r}
56 \\
-\ 25 \\
\hline 31
\end{array}
$$

_____

b.
$$
\begin{array}{r}
736 \\
-\ 514 \\
\hline 223
\end{array}
$$

_____

c.
$$
\begin{array}{r}
9614 \\
-\ 8312 \\
\hline 1312
\end{array}
$$

_____

Review these examples for Objective 5:

6. Subtract 29 from 46.

Write the problem vertically. In the ones column 6 is less than 9, so we must regroup 1 ten as 10 ones.

$$
\begin{array}{r}
\overset{3}{\cancel{4}}\ \overset{16}{\cancel{6}} \\
-\ 2\ 9
\end{array}
$$

Now subtract 9 from 16 ones in the ones column. Then subtract 2 tens from 3 tens in the tens column.

$$
\begin{array}{r}
\overset{3}{\cancel{4}}\ \overset{16}{\cancel{6}} \\
-\ 2\ 9 \\
\hline 1\ 7
\end{array}
$$

Finally, $46 - 29 = 17$. Check by adding 29 and 17; you should get 46.

Now Try:

6. Subtract 29 from 68.

_____

7. Subtract by regrouping when necessary.

645

− 449

Regroup 1 ten as 10 ones.

$$\begin{array}{r} \overset{3\ \ 15}{6\ \cancel{4}\ \cancel{5}} \\ -\ 4\ 4\ 9 \\ \hline 6 \end{array}$$

Need to regroup further because 3 is less than 4 in the tens column.

$$\begin{array}{r} \overset{5\ \ 13\ 15}{\cancel{6}\ \cancel{4}\ \cancel{5}} \\ -\ 4\ 4\ 9 \\ \hline 1\ 9\ 6 \end{array}$$

The difference is 196.

8. Subtract.

5203

− 2857

There are no tens that can be regrouped into ones. So you must first regroup 1 hundred as 10 tens.

$$\begin{array}{r} \overset{1\ \ 10}{5\ \cancel{2}\ \cancel{0}\ 3} \\ -\ 2\ 8\ 5\ 7 \end{array}$$

Now we may regroup from the tens position.

$$\begin{array}{r} \overset{1\ \overset{9}{\cancel{10}}\ 13}{5\ \cancel{2}\ \cancel{0}\ \cancel{3}} \\ -\ 2\ 8\ 5\ 7 \\ \hline 6 \end{array}$$

Complete the process.

$$\begin{array}{r} \overset{4\ \overset{11}{\cancel{1}}\ \overset{9}{\cancel{10}}\ 13}{\cancel{5}\ \cancel{2}\ \cancel{0}\ \cancel{3}} \\ -\ 2\ 8\ 5\ 7 \\ \hline 2\ 3\ 4\ 6 \end{array}$$

The difference is 2346.

9. Subtract.

6000

− 3888

There are no tens or hundreds that can be regrouped into ones. So you must first regroup 1 thousand as 10 hundreds, 1 hundred as 10 tens, and 1 ten as 10 ones.

---

7. Subtract by regrouping when necessary.

726

− 238

_____

8. Subtract.

4203

− 1415

_____

9. Subtract.

8000

− 2997

_____

$$\begin{array}{r} \overset{9}{} \;\; \overset{9}{} \\ 5 \; \cancel{10} \; \cancel{10} \; 10 \\ \cancel{6} \; \cancel{0} \; \cancel{0} \; \cancel{0} \\ -\;\; 3\; 8\; 8\; 8 \\ \hline 2\; 1\; 1\; 2 \end{array}$$

The difference is 2112.

10. Use addition to check each answer.

    a.
$$\begin{array}{r} 8\,2\,3 \\ -\;5\,8\,5 \\ \hline 2\,3\,8 \end{array}$$

Check.
$$\begin{array}{r} 5\,8\,5 \\ +\;2\,3\,8 \\ \hline 8\,2\,3 \end{array}$$

Correct.

    b.
$$\begin{array}{r} 5\,6\,1 \\ -\;2\,7\,3 \\ \hline 3\,0\,8 \end{array}$$

Check.
$$\begin{array}{r} 2\,7\,3 \\ +\;3\,0\,8 \\ \hline 5\,8\,1 \end{array}$$

Incorrect. Rework the original problem to get the correct answer.

$$\begin{array}{r} 4 \;\; 15\; 11 \\ \cancel{5}\; \cancel{6}\; \cancel{1} \\ -\; 2\; 7\; 3 \\ \hline 2\; 8\; 8 \end{array}$$

10. Use addition to check each answer.

    a.
$$\begin{array}{r} 7\,3\,8 \\ -\;2\,4\,9 \\ \hline 4\,8\,9 \end{array}$$

_____

    b.
$$\begin{array}{r} 6\,1\,4 \\ -\;2\,3\,6 \\ \hline 4\,8\,8 \end{array}$$

_____

Review this example for Objective 6:

11. Sally Tanner had $22,143 withheld from her paycheck last year for income tax. She actually owes only $16,959 in tax. What refund should she receive?

To find out the refund, subtract $16,959 from $22,143.

$$\begin{array}{r} 1\;\; 11\; 10\; 13\; 13 \\ \$\,\cancel{2}\;\cancel{2},\cancel{1}\;\cancel{4}\;\cancel{3} \\ -\; \$1\; 6,\, 9\; 5\; 9 \\ \hline \$5\; 1\; 8\; 4 \end{array}$$

The refund should be $5184.

Now Try:

11. An airplane is carrying 234 passengers. When it lands in Atlanta, 139 passengers get off the plane. How many passengers are then left on the plane?

_____

Objective 1   Change addition problems to subtraction problems or the reverse.

For extra help, see Examples 1–2 on pages 747–748 of your text and Section Lecture video for Section R.2.

Write two subtraction problems for each addition problem.

1. $149 + 38 = 187$

1. _____

2. $478 + 239 = 717$

2. _____

Write an addition problem for each subtraction problem.

3. $1211 - 426 = 785$

3. _____

Objective 2   Identify the minuend, subtrahend, and difference.

For extra help, see page 748 of your text and Section Lecture video for Section R.2.

Identify the minuend, subtrahend, and difference in each of the following subtraction problems.

4. $98 - 36 = 62$

4.

Minuend _____

Subtrahend _____

Difference _____

5. $35 - 9 = 24$

5.

Minuend _____

Subtrahend _____

Difference _____

Objective 3   Subtract when no regrouping is needed.

For extra help, see Example 3 on page 748 of your text and Section Lecture video for Section R.2.

Subtract.

6.   5573
          $-\ 422$

6. _____

7.   8539
          $-\ 2527$

7. _____

Objective 4   Use addition to check subtraction answers.

For extra help, see Example 4 on page 749 of your text and Section Lecture video for Section R.2.

Check the following subtractions. If an answer is not correct, give the correct answer.

8. $\begin{array}{r} 31{,}146 \\ -\ 7312 \\ \hline 23{,}834 \end{array}$ 8. _____

9. $\begin{array}{r} 192 \\ -\ 39 \\ \hline 167 \end{array}$ 9. _____

10. $\begin{array}{r} 4847 \\ -\ 3768 \\ \hline 1121 \end{array}$ 10. _____

Objective 5   Subtract with regrouping.

For extra help, see Examples 5–9 on pages 749–752 of your text and Section Lecture video for Section R.2.

Subtract.

11. $\begin{array}{r} 927 \\ -\ 729 \end{array}$ 11. _____

12. $\begin{array}{r} 33{,}728 \\ -\ \ 7829 \end{array}$ 12. _____

13. $\begin{array}{r} 7000 \\ -\ 297 \end{array}$ 13. _____

Objective 6   Use subtraction to solve application problems.

For extra help, see Example 10 on page 752 of your text and Section Lecture video for Section R.2.

Solve each application problem.

14.   On Sunday, 7342 people went to a football game, while on Monday, 9138 people went. How many more people went on Monday?   14. _____

15. The Conrads now pay $439 per month for rent. If they rent a larger apartment, the payment will be $702 per month. How much extra will they pay each month?

15. _____

16. One bid for painting a house was $2134. A second bid was $1954. How much would be saved using the second bid?

16. _____

Chapter R WHOLE NUMBERS REVIEW

R.3   Multiplying Whole Numbers

Learning Objectives	
1	Identify the parts of a multiplication problem.
2	Do chain multiplications.
3	Multiply by single-digit numbers.
4	Multiply quickly by numbers ending in zeros.
5	Multiply by numbers having more than one digit.
6	Use multiplication to solve application problems.

Key Terms

Use the vocabulary terms listed below to complete each statement in exercises 1−7.

factors   product   commutative property of multiplication

associative property of multiplication   chain multiplication problem

multiple   partial products

1.   By the_____, changing the order of the factors in a multiplication problem does not change the product.

2.   In multiplication, the numbers being multiplied are called the _____.

3.   By the_____, changing the grouping of the factors in a multiplication problem does not change the product.

4.   The product of two whole number factors is called a _____ of either factor.

5.   The answer to a multiplication problem is called the _____.

6.   The products found when multiplying numbers having two or more digits are called _____.

7.   A multiplication problem with more than two factors is a _____.

Guided Examples

Review these examples for Objective 1:
1.   Multiply.

   a. 3×7

   $3 \times 7 = 21$

Now Try:
1.   Multiply.

   a. 7×6

   _____

b. $9 \cdot 0$

$9 \cdot 0 = 0$

c. $(5)(9)$

$(5)(9) = 45$

b. $0 \cdot 5$

_____

c. $(2)(8)$

_____

Review this example for Objective 2:

2. Multiply $4 \times 3 \times 2$.

Parentheses show what to do first.
$(4 \times 3) \times 2$
$12 \times 2 = 24$
Also,
$4 \times (3 \times 2)$
$4 \times \quad 6 = 24$
Either grouping results in the same product.

Now Try:

2. Multiply $5 \times 4 \times 2$.

_____

Review these examples for Objective 3:

3. Multiply.

a. 6 4
$\times$ 3

Start by multiplying in the ones column.
$4 \times 3 = 12$ ones
 Write 1 ten in the tens column.
 Write 2 ones in the ones column.

  ¹
  6 4
$\times$ 3
─────
    2

Next, multiply 3 times 6 tens.
3×6 tens $= 18$ tens
Add the 1 ten that was written at the top of the tens column.
 18 tens + 1 ten = 19 tens

  ¹
  6 4
$\times$ 3
─────
1 9 2

b. 6 4 5
$\times$ 4

$5 \times 4 = 20$ ones; write 0 ones; write 2 tens in the tens column.

Now Try:

3. Multiply.

a. 4 3
$\times$ 6

_____

b. 5 6 7
$\times$ 7

_____

$$\begin{array}{r} \overset{2}{6\ 4\ 5} \\ \times 4 \\ \hline 0 \end{array}$$

$4 \times 4 = 16$ tens; add the 2 regrouped tens to get 18 tens; write 8 tens; write 1 hundred in the hundreds column.

$$\begin{array}{r} \overset{1\ 2}{6\ 4\ 5} \\ \times 4 \\ \hline 8\ 0 \end{array}$$

$4 \times 6 = 24$ hundreds; add the 1 regrouped hundred to get 25 hundreds.

$$\begin{array}{r} \overset{1\ 2}{6\ 4\ 5} \\ \times 4 \\ \hline 2\ 5\ 8\ 0 \end{array}$$

Review these examples for Objective 4:

4. Multiply.

   a. 67×10

   $67 \times 10 = 670$ Attach 0.

   b. 39×100

   $39 \times 100 = 3900$ Attach 00.

   c. 74×1000

   $74 \times 1000 = 74,000$ Attach 000.

5. Multiply.

   a. 67×4000

   Multiply 67 by 4, and then attach three zeros.

   $$\begin{array}{r} 6\ 7 \\ \times 4 \\ \hline 2\ 6\ 8 \end{array}$$

   $67 \times 4000 = 268,000$

   b. 170×80

   Multiply 17 by 8, and then attach two zeros.

   $$\begin{array}{r} 1\ 7 \\ \times 8 \\ \hline 1\ 3\ 6 \end{array}$$

   $170 \times 80 = 13,600$

Now Try:

4. Multiply.

   a. 46×10

   _____

   b. 708×100

   _____

   c. 519×1000

   _____

5. Multiply.

   a. 89×200

   _____

   b. 140×60

   _____

Review these examples for Objective 5:

6. Multiply.

a. $(53)(36)$

First write in vertical format. Then start by multiplying 53 by 6.

$$
\begin{array}{r}
\overset{1}{}5\ 3 \\
\times\ \ 6 \\
\hline
3\ 1\ 8
\end{array}
\quad \leftarrow 53 \times 6 = 318
$$

Now multiply 53 by 30.

$$
\begin{array}{r}
5\ 3 \\
\times\ \ 3\ 0 \\
\hline
1\ 5\ 9\ 0
\end{array}
\quad \leftarrow 53 \times 30 = 1590
$$

Add the results.

$$
\begin{array}{r}
5\ 3 \\
\times\ \ 3\ 6 \\
\hline
3\ 1\ 8 \quad \leftarrow 53 \times 6 \\
1\ 5\ 9\ 0 \quad \leftarrow 53 \times 30 \\
\hline
1\ 9\ 0\ 8
\end{array}
$$

Both 318 and 1590 are called partial products. As a common practice and to save time, the 0 in 1590 is usually not written.

$$
\begin{array}{r}
5\ 3 \\
\times\ \ 3\ 6 \\
\hline
3\ 1\ 8 \\
1\ 5\ 9 \\
\hline
1\ 9\ 0\ 8
\end{array}
$$

b.
$$
\begin{array}{r}
2\ 3\ 1 \\
\times\ \ 3\ 2\ 2 \\
\hline
\end{array}
$$

Be sure to align numbers in columns of partial products

$$
\begin{array}{r}
2\ 3\ 1 \\
\times\ \ 3\ 2\ 2 \\
\hline
4\ 6\ 2 \\
4\ 6\ 2 \\
6\ 9\ 3 \\
\hline
7\ 4,3\ 8\ 2
\end{array}
$$

c.
$$
\begin{array}{r}
6\ 4\ 7 \\
\times\ \ 5\ 4 \\
\hline
\end{array}
$$

First, multiply by 4.

Now Try:

6. Multiply.

a. $(45)(64)$

_____

b.
$$
\begin{array}{r}
1\ 3\ 2 \\
\times\ \ 2\ 1\ 3 \\
\hline
\end{array}
$$

_____

c.
$$
\begin{array}{r}
1\ 6\ 5 \\
\times\ \ 5\ 3 \\
\hline
\end{array}
$$

_____

```
  1 2
  6 4 7
×   5 4
-------
2 5 8 8
```

Now multiply by 5, being careful to line up the tens. Finally, add the partial products.

```
    2 3
    1 2
  6 4 7
×     5 4
---------
  2 5 8 8
  3 2 3 5
---------
3 4, 9 3 8
```

7. Multiply.

  a.   1 4 6

     × 2 0 5

Align numbers in columns of partial products

```
    1 4 6
×   2 0 5
---------
    7 3 0
  0 0 0
2 9 2
---------
2 9, 9 3 0
```

  b.   1 0 6 2

     × 3 0 0 4

Be careful to line up the zeros.

```
    1 0 6 2      or        1 0 6 2
×     3 0 0 4          ×     3 0 0 4
-----------           -----------
      4 2 4 8               4 2 4 8
    0 0 0 0             3 1 8 6 0 0
  0 0 0 0              -----------
3 1 8 6                3, 1 9 0, 2 4 8
-----------
3, 1 9 0, 2 4 8
```

7. Multiply.

  a.   8 4 3

     × 1 0 6

_____

  b.   4 6 2 5

     × 7 0 0 3

_____

Review this example for Objective 6:

8. Find the total cost of 85 tickets at $29 each.

Multiply 85 by 29.

```
    8 5
×   2 9
-------
  7 6 5
1 7 0
-------
2 4 6 5
```

The total cost of the tickets is $2465.

Now Try:

8. Find the total cost of 316 pair of gloves at $17 per pair.

_____

Objective 1   Identify the parts of a multiplication problem.

For extra help, see Example 1 on page 755 of your text and Section Lecture video for Section R.3.

Identify the factors and the product in each multiplication problem.

1. $108 = 9 \times 12$

     1.

     Factors _____

     Product _____

Multiply.

2. $4 \cdot 6$

     2. _____

3. 8×9

     3. _____

Objective 2   Do chain multiplications.

For extra help, see Example 2 on page 756 of your text and Section Lecture video for Section R.3.

Multiply.

4. $4 \times 4 \times 2$

     4. _____

5. $3 \times 4 \times 7$

     5. _____

6. $(6)(4)(8)$

     6. _____

Objective 3   Multiply by single-digit numbers.

For extra help, see Example 3 on pages 756–757 of your text and Section Lecture video for Section R.3.

Multiply.

7. $\begin{array}{r} 54 \\ \times\ 4 \\ \hline \end{array}$

     7. _____

8. $\begin{array}{r} 163 \\ \times\ \ 5 \\ \hline \end{array}$

     8. _____

9.   30,009
         ×   6

9. _____

Objective 4   Multiply quickly by numbers ending in zeros.

For extra help, see Examples 4–5 on pages 757–758 of your text and Section Lecture video for Section R.3.

Multiply.

10.   (852)(30)

10. _____

11. 500×40

11. _____

12. 8234×2000

12. _____

Objective 5   Multiply by numbers having more than one digit.

For extra help, see Examples 6–7 on pages 758–759 of your text and Section Lecture video for Section R.3.

Multiply.

13.   9 2
           ×   4 8

13. _____

14.   6 4 4
           ×   1 9

14. _____

15.   4 0 3 1
        ×   4 0 8

15. _____

Objective 6   **Use multiplication to solve application problems.**

For extra help, see Example 8 on page 760 of your text and Section Lecture video for Section R.3.

Solve the following application problems.

16. On a recent trip the Jensen family drove 45 miles per hour on the average. They drove 22 hours altogether. How many miles did they drive altogether?

16. _____

Find the total cost of the following items.

17. 512 boxes of chalk at $19 per box

17. _____

18. 178 baseball caps at $9 per cap

18. _____

Chapter R WHOLE NUMBERS REVIEW

R.4   Dividing Whole Numbers

Learning Objectives
1   Write division problems in three ways.
2   Identify the parts of a division problem.
3   Divide 0 by a number.
4   Recognize that division by 0 is undefined.
5   Divide a number by itself.
6   Use short division.
7   Use multiplication to check quotients.
8   Use tests for divisibility.

Key Terms

Use the vocabulary terms listed below to complete each statement in exercises 1−5.

dividend   divisor   quotient   short division   remainder

1.   The number left over when two numbers do not divide exactly is the

_____.

2.   The number being divided by another number in a division problem is the

_____.

3.   The answer to a division problem is called the _____.

4.   In the problem $639 \div 9$, 9 is called the _____.

5.   _____ is a method of dividing a number by a one-digit divisor.

Guided Examples

Review these examples for Objective 1:

1.   Write each division problem using two other symbols.

   a. $24 \div 8 = 3$

   $8\overline{)24}$ = 3   or $\dfrac{24}{8} = 3$

   b. $\dfrac{27}{3} = 9$

   $27 \div 3 = 9$ or $3\overline{)27}$ = 9

Now Try:

1.   Write each division problem using two other symbols.

   a. $28 \div 4 = 7$

   b. $\dfrac{72}{9} = 8$

   _____

c. $6\overline{)30}^{\,5}$

$$30 \div 6 = 5 \quad \text{or} \quad \frac{30}{6} = 5$$

c. $8\overline{)32}^{\,4}$

_____

Review these examples for Objective 2:

2. Identify the dividend, divisor and quotient.

a. $45 \div 5 = 9$

$$45 \div 5 = 9 \;\leftarrow \text{Quotient}$$
$$\nearrow \qquad \nwarrow$$
$$\text{Dividend} \qquad \text{Divisor}$$

b. $\dfrac{90}{30} = 3$

Dividend
$\downarrow$
$\dfrac{90}{30} = 3 \leftarrow$ Quotient
$\uparrow$
Divisor

c. $8\overline{)48}^{\,6}$

$6 \leftarrow$ Quotient
$8\overline{)48} \leftarrow$ Dividend
$\uparrow$
Divisor

Now Try:

2. Identify the dividend, divisor and quotient.

a. $63 \div 9 = 7$

_____

b. $\dfrac{56}{8} = 7$

_____

c. $6\overline{)54}^{\,9}$

_____

Review these examples for Objective 3:

3. Divide.

a. $0 \div 17$

$0 \div 17 = 0$

b. $0 \div 9674$

$0 \div 9674 = 0$

c. $\dfrac{0}{668}$

$\dfrac{0}{668} = 0$

d. $139\overline{)0}$

$139\overline{)0}^{\,0}$

Now Try:

3. Divide.

a. $0 \div 6$

_____

b. $0 \div 1283$

_____

c. $\dfrac{0}{25}$

_____

d. $67\overline{)0}$

_____

4. Change each division problem to a multiplication problem.

    a. $\dfrac{24}{3} = 8$

        $\dfrac{24}{3} = 8$ becomes $3 \cdot 8 = 24$ or $8 \cdot 3 = 24$

    b. $7\overline{)42}$ with quotient 6

        $7\overline{)42}$ becomes $6 \cdot 7 = 42$ or $7 \cdot 6 = 42$

    c. $63 \div 7 = 9$

        $63 \div 7 = 9$ becomes $7 \cdot 9 = 63$ or $9 \cdot 7 = 63$

4. Change each division problem to a multiplication problem.

    a. $\dfrac{36}{4} = 9$

        _____

    b. $3\overline{)18}$ with quotient 6

        _____

    c. $40 \div 8 = 5$

        _____

Review these examples for Objective 4:

5. Find the quotient whenever possible.

    a. $\dfrac{13}{0}$

        $\dfrac{13}{0}$ is undefined.

    b. $0\overline{)21}$

        $0\overline{)21}$ is undefined.

    c. $27 \div 0$

        $27 \div 0$ is undefined.

Now Try:

5. Find the quotient whenever possible.

    a. $\dfrac{9}{0}$

        _____

    b. $0\overline{)21}$

        _____

    c. $110 \div 0$

        _____

Review these examples for Objective 5:

6. Divide.

    a. $18 \div 18$

        $18 \div 18 = 1$

    b. $42\overline{)42}$

        $42\overline{)42}$ with quotient 1

    c. $\dfrac{95}{95}$

        $\dfrac{95}{95} = 1$

Now Try:

6. Divide.

    a. $7 \div 7$

        _____

    b. $19\overline{)19}$

        _____

    c. $\dfrac{89}{89}$

        _____

Review these examples for Objective 6:

7. Divide: $4\overline{)84}$.

First, divide 8 by 4.

$$4\overline{)\overset{2}{84}} \qquad \frac{8}{4}=2$$

Next, divide 4 by 4.

$$4\overline{)\overset{21}{84}} \qquad \frac{4}{4}=1$$

8. Divide 167 by 3.

$$3\overline{)167}$$

Because 1 cannot be divided by 3, divide 16 by 3. Notice that the 5 is placed over the 6 in 16.

$$3\overline{)\overset{5}{16^17}} \qquad \frac{16}{3}=5 \text{ with 1 left over}$$

Next divide 17 by 3. The final number left over is the remainder.

$$3\overline{)\overset{5\ 5}{16^17}} \text{ R2} \qquad \frac{17}{3}=5 \text{ with 2 left over}$$

9. Divide 1943 by 8.

Divide 19 by 8.

$$8\overline{)\overset{2}{19^343}} \qquad \frac{19}{8}=2 \text{ with 3 left over}$$

Divide 34 by 8.

$$8\overline{)\overset{2\ 4}{19^34^23}} \qquad \frac{34}{8}=4 \text{ with 2 left over}$$

Divide 23 by 8.

$$8\overline{)\overset{2\ 4\ 2}{19^34^23}} \text{ R7} \qquad \frac{23}{8}=2 \text{ with 7 left over}$$

The quotient is 242 R7.

Review these examples for Objective 7:

10. Check each answer.

a. $6\overline{)\overset{91}{549}}$ R3

To check, multiply the divisor and the quotient, then add the remainder to get the dividend.

Now Try:

7. Divide: $3\overline{)69}$.

_____

8. Divide 149 by 2.

_____

9. Divide 2354 by 9.

_____

Now Try:

10. Check each answer.

a. $3\overline{)\overset{28}{85}}$ R1

_____

$$(6 \times 91) + 3$$
$$546 \ + 3 = 549$$

549 matches the original dividend, so the division was done correctly.

b. $8 \overline{)2315}$ $\dfrac{288}{}$ R6

To check, multiply the divisor and the quotient, then add the remainder to get the dividend.

$$(8 \times 288) + 6$$
$$2304 \ + 6 = 2310$$

The answer does not check. Rework the original problem to get the correct answer, 289 R 3.

b. $6 \overline{)469}$ $\dfrac{79}{}$ R4

_____

Review these examples for Objective 8:

11. Are the following numbers divisible by 2?

a. 864

Because the number ends in 4, which is an even number, the number 864 is divisible by 2.

b. 4287

4287 is not divisible by 2.
Ends in 7, not 0, 2, 4, 6, or 8

12. Are the following numbers divisible by 3?

a. 537

Add the digits.
$$5 + 3 + 7 = 15$$
Because 15 is divisible by 3, the number 537 is also divisible by 3.

b. 156,304

Add the digits.
$$1 + 5 + 6 + 3 + 0 + 4 = 19$$
Because 19 is not divisible by 3, the number 156,304 is not divisible by 3.

13. Are the following numbers divisible by 5?

a. 43,760

43,760 ends in 0 and is divisible by 5.

b. 7565

7565 ends in 5 and is divisible by 5.

Now Try:

11. Are the following numbers divisible by 2?

a. 568

_____

b. 873

_____

12. Are the following numbers divisible by 3?

a. 843

_____

b. 386,452

_____

13. Are the following numbers divisible by 5?

a. 290

_____

b. 970,405

_____

c. 954

954 ends in 4 and is not divisible by 5.

14. Are the following numbers divisible by 10?

a. 950 and 8790

950 and 8790 both end in 0 and are divisible by 10.

b. 366 and 27,853

366 and 27,853 do not end in 0 and are not divisible by 10.

c. 7356

_____

14. Are the following numbers divisible by 10?

a. 360

_____

b. 565

_____

Objective 1   Write division problems in three ways.

For extra help, see Example 1 on page 763 of your text and Section Lecture video for Section R.4.

Write each division problem using two other symbols.

1. $15 \div 3 = 5$

1. _____

2. $\dfrac{50}{25} = 2$

2. _____

Objective 2   Identify the parts of a division problem.

For extra help, see Example 2 on pages 763–764 of your text and Section Lecture video for Section R.4.

Identify the dividend, divisor, and quotient.

3. $63 \div 7 = 9$

3.

Dividend _____

Divisor _____

Quotient _____

4. $5\overline{)30}$ (quotient 6)

4.

Dividend _____

Divisor _____

Quotient _____

5. $\dfrac{44}{11} = 4$

5.
Dividend _____

Divisor _____

Quotient _____

Objective 3   Divide 0 by a number.

For extra help, see Examples 3–4 on page 764 of your text and Section Lecture video for Section R.4.

Divide.

6. $12\overline{)0}$

6. _____

7. $\dfrac{0}{6}$

7. _____

8. $0 \div 15$

8. _____

Objective 4   Recognize that division by 0 is undefined.

For extra help, see Example 5 on page 765 of your text and Section Lecture video for Section R.4.

Divide. If the division is not possible, write "undefined."

9. $0\overline{)72}$

9. _____

10. $\dfrac{7}{0}$

10. _____

11. $9 \div 0$

11. _____

Objective 5   Divide a number by itself.

For extra help, see Example 6 on page 766 of your text and Section Lecture video for Section R.4.

Divide.

12. $77 \div 77$

12. _____

13. $\dfrac{421}{421}$

13. _____

Objective 6   Use short division.

For extra help, see Examples 7–9 on pages 766–767 of your text and Section Lecture video for Section R.4.

Divide by using short division.

14. $724 \div 5$ **14.** _____

15. $\dfrac{651}{9}$ **15.** _____

16. $8\overline{)1135}$ **16.** _____

Objective 7   Use multiplication to check quotients.

For extra help, see Example 10 on pages 767–768 of your text and Section Lecture video for Section R.4.

Use multiplication to check each answer. If an answer is incorrect, find the correct answer.

17. $6\overline{)9137}$ 1522 R4   **17.** _____

18. $3852 \div 4 = 963$ **18.** _____

19. $\dfrac{18{,}150}{3} = 650$ **19.** _____

Objective 8   Use tests for divisibility.

For extra help, see Examples 11–14 on pages 769–770 of your text and Section Lecture video for Section R.4.

Determine if the following numbers are divisible by 2, 3, 5, or 10 Write **yes** *or* **no**.

20.   897   **20. 2:** _____

                                                          3: _____

                                                          5: _____

                                                          10: _____

Name: Date:
Instructor: Section:

21.   908   **21.** **2:** _____

                                                  3: _____

                                                  5: _____

                                                  10: _____

22.   6205   **22.** **2:** _____

                                                  3: _____

                                                  5: _____

                                                  10: _____

Chapter R WHOLE NUMBERS REVIEW

R.5 Long Division

Learning Objectives
1   Use long division.
2   Divide by multiples of 10.
3   Use multiplication to check quotients.

Key Terms

Use the vocabulary terms listed below to complete each statement in exercises 1–5.

      long division   **dividend**   **divisor**   **quotient**   **remainder**

1. In the problem $751 \div 23 = 32 \text{ R } 15$, 751 is called the _____.

2. In the problem $751 \div 23 = 32 \text{ R } 15$, 15 is called the _____.

3. In the problem $751 \div 23 = 32 \text{ R } 15$, 23 is called the _____.

4. In the problem $751 \div 23 = 32 \text{ R } 15$, 32 is called the _____.

5. _____ is a method of dividing a number by a divisor with more than one digit.

Guided Examples

Review these examples for Objective 1:   **Now Try:**

1. Divide. $62\overline{)5146}$ 1. Divide. $34\overline{)2482}$

     Because 62 is closer to 60 than 70, use the first digit of the divisor as a trial divisor.
     Try to divide the first digit of the dividend by 6. Since 5 cannot be divided by 6, use the first two digits, 51.

$$\frac{51}{6} = 8 \text{ R}3$$

$$62\overline{)5146}^{8}$$

     Multiply 8 and 62 to get 496; next, subtract 496 from 514.

$$\begin{array}{r} 8 \\ 62\overline{)5146} \\ \underline{496} \\ 18 \end{array}$$

Bring down the 6 on the right.

$$62\overline{)5146}$$
with 8 on top
$$\underline{496}$$
$$186$$

Use the trial divisor 6. $\dfrac{18}{6} = 3$

$$62\overline{)5146}$$
with 83 on top
$$\underline{496}$$
$$186$$
$$\underline{186}$$
$$0$$

Check: $62 \times 83 = 5146$

2. Divide. $78\overline{)4139}$

Use 8 as a trial divisor, since 78 is closer to 80 than 70.

$\dfrac{41}{8} = 5$ with 1 left over

$$78\overline{)4139}$$
with 5 on top
$$\underline{390} \quad \leftarrow 5 \times 78 = 390$$
$$23 \quad \leftarrow 413 - 390 = 23 \text{ (smaller}$$
$$\text{than 78, the divisor)}$$

Bring down the 9.

$$78\overline{)4139}$$
with 5 on top
$$\underline{390}$$
$$239$$

$\dfrac{23}{8} = 2$ with 7 left over

$$78\overline{)4139}$$
with 52 on top
$$\underline{390}$$
$$239$$
$$\underline{156} \quad \leftarrow 2 \times 78 = 156$$
$$83 \quad \leftarrow \text{Greater than 78}$$

The remainder 83 is greater than the divisor 78, so 3 should be used instead of 2.

2. Divide. $59\overline{)2297}$

_____

$$\begin{array}{r}53 \text{ R}5 \\ 78\overline{)4139} \\ 390 \\ \hline 239 \\ 234 \\ \hline 5\end{array}$$ $\quad \leftarrow 3 \times 78 = 234$

3. Divide. $27\overline{)8340}$

$$\begin{array}{r}3 \\ 27\overline{)8340} \\ 81 \\ \hline 2\end{array}$$ $\quad \leftarrow 3 \times 27 = 81$
$\qquad\qquad \leftarrow 83 - 81 = 2$

Bring down the 4.

$$\begin{array}{r}3 \\ 27\overline{)8340} \\ 81 \\ \hline 24\end{array}$$

Since 24 cannot be divided by 27, write 0 in the quotient as a placeholder.

$$\begin{array}{r}30 \\ 27\overline{)8340} \\ 81 \\ \hline 24\end{array}$$

Bring down the final digit, the 0.

$$\begin{array}{r}30 \\ 27\overline{)8340} \\ 81 \\ \hline 240\end{array}$$

Complete the problem.

$$\begin{array}{r}308 \text{ R}24 \\ 27\overline{)8340} \\ 81 \\ \hline 240 \\ 216 \\ \hline 24\end{array}$$

The quotient is 308 R24.

Review these examples for Objective 2:

4. Divide:

a. $40 \div 10$

Drop one 0 in divisor and dividend.
$40 \div 10 = 4$

---

3. Divide. $26\overline{)5454}$

_____

Now Try:

4. Divide:

a. $50 \div 10$

_____

b. $6500 \div 100$

Drop two zeros in divisor and dividend.
$6500 \div 100 = 65$

c. $865,000 \div 1000$

Drop three zeros in divisor and dividend.
$865,000 \div 1000 = 865$

5. Divide:

a. $30\overline{)11,250}$

Drop one zero from the divisor and the dividend.

$$\begin{array}{r} 375 \\ 3\overline{)1125} \\ \underline{9} \\ 22 \\ \underline{21} \\ 15 \\ \underline{15} \\ 0 \end{array}$$

Since $1125 \div 3$ is 375, then $11,250 \div 30$ is 375.

b. $5400\overline{)37,800}$

Drop two zeros from the divisor and the dividend.

$$\begin{array}{r} 7 \\ 54\overline{)378} \\ \underline{378} \\ 0 \end{array}$$

Since $378 \div 54$ is 7, then $37,800 \div 5400$ is 7.

Review this example for Objective 3:

6. Check the quotient:

$$\begin{array}{r} 34\,\text{R}120 \\ 415\overline{)15,890} \end{array}$$

Multiply the quotient and the divisor. Then add the remainder.

$$\begin{array}{r} 4\ 1\ 5 \\ \times\quad 3\ 4 \\ \hline 1\ 6\ 6\ 0 \\ 1\ 2\ 4\ 5 \\ \hline 1\ 4\ 1\ 1\ 0 \\ +\quad 1\ 2\ 0 \\ \hline 1\ 4,2\ 3\ 0 \end{array}$$

The answer does not check. Rework the original problem to get 38 R120.

b. $4700 \div 100$

_____

c. $606,000 \div 1000$

_____

5. Divide:

a. $60\overline{)8100}$

_____

b. $160\overline{)320,960}$

_____

Now Try:

6. Check the quotient:

$$\begin{array}{r} 36\,\text{R}62 \\ 532\overline{)19,746} \end{array}$$

_____

Name: Date:

Instructor: Section:

Objective 1   Use long division.

For extra help, see Examples 1–3 on pages 773–775 of your text and Section Lecture video for Section R.5.

Divide using long division. Check each answer.

1. $42\overline{)3234}$ 1. _____

2. $94\overline{)29,047}$ 2. _____

3. $657\overline{)429,700}$ 3. _____

Objective 2   Divide by multiples of 10.

For extra help, see Examples 4–5 on pages 775–776 of your text and Section Lecture video for Section R.5.

Divide.

4. $80\overline{)560}$ 4. _____

5. $800\overline{)10,400}$ 5. _____

6. $750\overline{)25,500}$ 6. _____

Objective 3   Use multiplication to check division answers.

For extra help, see Example 6 on page 776 of your text and Section Lecture video for Section R.5.

Check each answer. If an answer is incorrect, give the correct answer.

7. $89\overline{)5790}$ 65 R 5   7. _____

8. $103\overline{)4658}$ 44 R 22   8. _____

9. $428\overline{)196{,}883}$ 400 R 30   9. _____

Chapter 1 INTRODUCTION TO ALGEBRA: INTEGERS

1.1 Place Value

Key Terms

1. digits

2. whole numbers

3. place value system

Now Try

1. 0, 5, 59, 350

2. hundred-thousands and hundreds

3a. nine million, seventy-five thousand, eight hundred sixty-two

3b. fifty-four trillion, eight hundred billion, five hundred forty-three million, seven hundred thousand, one hundred

4a. 753,006

4b. 11,000,010,025

Objective 1

1. 2

3. 1, 365

Objective 2

5. ten-millions

Objective 3

7. fifty-nine billion, five hundred four million, eight hundred six thousand, eight hundred seventy-three

9. 987, 000,330

1.2 Introduction to Integers

Key Terms

1. absolute value

2. number line

3. integers

Now Try

1a. −150 feet

1b. +68°F or 68°F

2.

3a. 0 > −8

3b. −3 > −9

3c. 11 > −4

4a. 12

4b. 13

4c. 0

Objective 1

1. −46

3. −1,200,000

Answers

Objective 2

5.
$$-5\ -4\ -3\ -2\ -1\ \ 0\ \ 1\ \ 2\ \ 3\ \ 4\ \ 5$$

Objective 3

7. $>$ 9. $<$

Objective 4

11. 21

1.3   Adding Integers

Key Terms

1. commutative property of addition   2. addends

3. associative property of addition   4. sum

5. addition property of 0

Now Try

1. -5 2a. -36 2b. 40

3a. -16 3b. 21   4. $-3°$

5a. $63+254=317$ 5b. $-77+(-44)=-121$

6a. $-20+(19+(-19))=-20+0=-20$

6b. $8+2+(-17)=(8+2)+(-17)=10+(-17)=-7$

Objective 1

1. -3 3. 18 points

Objective 2

5. $8+(-15+(-5))=-12$

1.4 Subtracting Integers

Key Terms

1. opposite   2. additive inverse

Now Try

1a. -25, $25+(-25)=0$ 1b. 39, $-39+39=0$ 1c. 0, $0+0=0$

2a. $13+(-15)=-2$ 2b. $-10+9=-1$ 2c. $12+11=23$

2d. $-8+(-16)=-24$ 3. -38

Objective 1

1. 11; $-11+11=0$ 3. 5; $-5+5=0$

Objective 2

5. -14

Objective 3

7. 0 9. -22

1.5 Problem Solving: Rounding and Estimating

Key Terms

1. front end rounding   2. rounding   3. estimate

Now Try

1a. $-\underline{3}2$; -30 1b. $\$\underline{6}42$; $\$600$ 1c. $-7\underline{6},542$; $-77,000$

2. 700 3. $95,000$ 4a. -4740

4b. $57,000$ 5a. $-80,000$ 5b. $246,000,000$

6a. -800 6b. $1,000,000$ 7. $\$2200$; $\$2412$

Objective 1

1. 3 3. 4

Objective 2

5. $810,000$

Objective 3

7. $200,000$ 9. $30°$; $35°$

1.6   Multiplying Integers

Key Terms

1. commutative property of multiplication   2. factors

3. associative property of multiplication

4. multiplication property of 0   5. product

6. distributive property   7. multiplication property of 1

Now Try

1a. $80 \cdot 6$ or $80(6)$ or $(80)(6)$. The factors are 80 and 6. The product is 480.

1b. $8 \cdot 15$ or $8(15)$ or $(8)(15)$. The factors are 8 and 15. The product is 120.

2a. -42 2b. 60   2c. -54

3a. -70 3b. -64

4a. 0; multiplication property of 0

4b. -92; multiplication property of 1

5a. $63 = 63$; commutative property of multiplication

5b. $-196 = -196$; associative property of multiplication

6a. $7(4) + 7(8)$; both results are 84

6b. $-9(-5) + (-9)(3)$; both results are 18

7. -1800; -1785

Objective 1

1. $-3 \cdot 6$; $(-3)(6)$ 3. $4 \cdot 5$; $(4)(5)$

Objective 2

5. -62

Objective 3

7. 0   9. $3 \cdot \big((-2) \cdot (-9)\big) = 54$

Objective 4

11. \$1300; \$1315

1.7   Dividing Integers

Key Terms

1. factors   2. quotient   3. product

Now Try

1a. −5   1b. 4   1c. −3

2a. 1 Any nonzero number divided by itself is 1.

2b. 93 Any number divided by 1 is the number.

2c. 0 Zero divided by any nonzero number is 0.

2d. undefined Division by 0 is undefined.   3a. 10

3b. −4   3c. −1   4. Est: $300; Exact: $309

5a. 6 packages with 2 hot dogs left without rolls, or 7 packages with 6 extra rolls

5b. 9 trips, 8 elevator trips would leave 6 people walking to the top

Objective 1

1. −7   3. −10

Objective 2

5. 1. Any nonzero number divided by itself is 1.

Objective 3

7. 2   9. −21

Objective 4

11. $23,000; $22,600

Objective 5

13. $96 \div 7 = 13$ R5. Every truck has at least 13 construction workers. The remainder of 5 means that 5 trucks each carry a 14th worker.

15. $1500 \div 468 = 3$ R 96. This means that each student receives three tickets, and there are 96 seats that are available for assignment.

1.8 Exponents and Order of Operations

Key Terms

1. order of operations   2. exponent

Now Try

1a. 5^4; 625; 5 to the fourth power

1b. 6^2; 36; 6 squared, or 6 to the second power

1c. 23^1; 23; 23 to the first power   2a. 49

2b. −1000   2c. 81   2d. −432

3a. −26   3b. 24   4. 13

5a. −24   5b. −42   6a. −17

6b. −53   7. −9

Objective 1

1. 12^3; twelve cubed   3. 7^7; seven to the seventh power

Objective 2

5. 144

Objective 3

7. 17   9. −16

Objective 4

11. 16

Chapter 2 UNDERSTANDING VARIABLES AND SOLVING EQUATIONS

2.1   Introduction to Variables

Key Terms

1. expression
2. variable
3. evaluate the expression
4. constant
5. coefficient

Now Try

1. Variable: c; constant: 5; $c + 5$

2a. 29; Order 29 lunches.

2b. 50; Order 50 lunches

3. 32 yards

4. 128

5a. 96

5b.
$10 + 21$ is 31	$7 \cdot 6$ is 42
$-12 + 9$ is -3	$-9 \cdot 5$ is -45
$93 + 0$ is 93	$31 \cdot 0$ is 0

6. $\dfrac{0}{b} = 0$

7a. $z \cdot z \cdot z \cdot z$

7b. $14 \cdot r \cdot s \cdot s$

7c. $-13 \cdot m \cdot m \cdot m \cdot n \cdot n \cdot n \cdot n \cdot n$

8a. 81

8b. 144

8c. -240

Objective 1

1. -7: constant; h: variable

3. 9: coefficient; k: variable; 1: constant

Objective 2

5. (a) $10,040 (b) $12,620

Objective 3

7. Multiplication is distributive over addition.

9. Zero added to any number equals that number.

Objective 4

11. $u \cdot u \cdot u \cdot v \cdot v \cdot w \cdot w$

2.2   Simplifying Expressions

Key Terms

1. like terms
2. term
3. simplify an expression
4. variable term
5. constant term

Now Try

1a. $-8a$ and a; -8 and 1 1b. $21r^2s$ and $-5r^2s$; 21 and -5

1c. 26 and -2; they are constants.   2a. $20x$

2b. $-39z^2$ 3a. $11pq + q$ 3b. $7c - 9$

4a. $108b$ 4b. $-36x$ 4c. $40y^2$

5a. $30x + 12$ 5b. $-54x - 36$ 5c. $28x - 112$

6. $6x - 6$

Objective 1

1. like terms: $2x$, $-3x$; coefficients: 2, -3 3. $-19c^3z^4$

Objective 2

5. $-6s - 5st - 16t + 8$

Objective 3

7. $-5t - 20$ 9. $20q - 3$

2.3 Solving Equations Using Addition

Key Terms

1. solution

2. addition property of equality

3. equation

4. solve an equation

5. check the solution

Now Try

1. 29

2a. $c = 16$

2b. $x = -2$

3a. $x = -11$

3b. $b = 6$

Objective 1

1. 12

3. 7

Objective 2

5. $n = -66$

Objective 3

7. $a = 3$

9. $n = -44$

2.4 Solving Equations Using Division

Key Terms

1. division property of equality

2. addition property of equality

Now Try

1a. $s = 6$

1b. $w = -16$

2a. $y = 4$

2b. $h = 0$

3. $y = -13$

Objective 1

1. $m = -7$

3. $h = 28$

Objective 2

5. $m = 33$

Objective 3

7. $b = -19$

9. $v = -4$

Answers

2.5   Solving Equations with Several Steps

Key Terms

1. addition property of equality

2. distributive property

3. division property of equality

Now Try

1. $n = 7$ 2. $x = -10$ 3. $m = 0$

4. $m = -5$

Objective 1

1. $m = 0$ 3. $q = -3$

Objective 2

5. $t = 6$

Chapter 3 SOLVING APPLICATION PROBLEMS

3.1   Problem Solving: Perimeter

Key Terms

1. triangle
2. perimeter
3. formula
4. square
5. rectangle
6. parallelogram

Now Try

1. 60 in.
2. 125 cm
3. 184 ft
4. 18 in.
5. 74 m
6. 70 mm
7. 306 in.

Objective 1

1. 28 ft
3. 19 mi

Objective 2

5. 13 ft

Objective 3

7. 22 in.
9. 255 cm

3.2   Problem Solving: Area

Key Terms

1. square

2. parallelogram

3. rectangle

4. area

Now Try

1a. 253 m^2

1b. 189 cm^2

2. 5 ft

3. 169 in.^2

4. 5 mi

5a. 171 cm^2

5b. 425 m^2

6. 9 ft

7. $1688 \text{ m}; 169,984 \text{ m}^2$

Objective 1

1. 10 cm^2

3. 76 cm

Objective 2

5. 2025 cm^2

Objective 3

7. 208 m^2

9. 6 m

Objective 4

11. 8 ft

3.3   Solving Application Problems with One Unknown Quantity

Key Terms

1. sum; increased by

2. product; double

3. quotient; per

4. difference; less than

Now Try

1a. $x + 7$ or $7 + x$

1b. $43 + x$ or $x + 43$

1c. $x + 63$ or $63 + x$

1d. $-20 + x$ or $x + (-20)$

1e. $x + 75$ or $75 + x$

1f. $x - 44$

1g. $89 - x$

1h. $x - 89$

1i. $x - 16$

1j. $x - 36$

1k. $48 - x$

2a. $57x$

2b. $43x$

2c. $2x$

2d. $\dfrac{-13}{x}$

2e. $\dfrac{x}{19}$

2f. $6x - 81$

3. $7x + 21 = 56$; $x = 5$

4. 3 pounds

5. 55 paper plates

6. $14.00

Objective 1

1. $-6x$

3. $1 + 3x$ or $3x + 1$

Objective 2

5. $3 + 7x = 31$; $x = 4$

Objective 3

7. 12 bananas

9. 7 celery sticks

3.4   Solving Application Problems with Two Unknown Quantities

Key Terms

1. added to; more than

2. times; triple

3. divided by; half

4. subtracted from; minus

Now Try

1. cat: $55; dog: $165

2. 38 m, 60 m

3. length: 31 in.; width: 12 in.

Objective 1

1. Greer: 8135 votes; Jones: 8651 votes

3. length: 36 yd; width: 12 yd

Chapter 4 RATIONAL NUMBERS: POSITIVE AND NEGATIVE FRACTIONS

4.1 Introduction to Signed Fractions

Key Terms

1. equivalent fractions　　2. fraction　　　　3. improper fraction

4. numerator　　　　　　5. proper fraction　　6. denominator

Now Try

1. shaded: $\frac{3}{7}$; unshaded: $\frac{4}{7}$　　　　　2. $\frac{9}{8}$

3a. Numerator: 5; denominator: 17; 17 equal parts in the whole

3b. Numerator: 6; denominator: 23; 23 equal parts in the whole

4a. $\frac{7}{8}$, $\frac{9}{11}$, $\frac{12}{29}$, $\frac{1}{5}$　　　4b. $\frac{22}{7}$, $\frac{13}{13}$, $\frac{17}{4}$　　　5.

6. $\frac{9}{10}$, $\frac{9}{10}$　　　　7a. $-\frac{24}{30}$　　　7b. $\frac{5}{6}$

8a. 1　　　　　　8b. –6　　　　　　8c. 13

Objective 1

1. $\frac{3}{5}$; $\frac{2}{5}$　　　　　3. $\frac{3}{5}$; $\frac{7}{5}$

Objective 2

5. 12; 5

Objective 3

7. 　　　　　　　9.

Objective 4

11. $\frac{8}{9}$

Objective 5

13. $-\frac{32}{48}$　　　　　15. $-\frac{1}{5}$

4.2   Writing Fractions in Lowest Terms

Key Terms

1. composite number   2. prime factorization   3. prime number

4. lowest terms

Now Try

1a. yes   1b. no; 9   2a. $\dfrac{5}{9}$

2b. $\dfrac{7}{8}$ 2c. $-\dfrac{5}{13}$ 2d. $\dfrac{11}{12}$

3. Prime: 29 and 31; composite: 8 and 21; neither: 1   4a. $2 \cdot 2 \cdot 2 \cdot 2 \cdot 2 \cdot 3$

4b. $2 \cdot 2 \cdot 2 \cdot 7$ 5a. $2 \cdot 2 \cdot 2 \cdot 19$ 5b. $2 \cdot 2 \cdot 2 \cdot 3 \cdot 7$

6a. $\dfrac{1}{3}$ 6b. $\dfrac{3}{10}$ 6c. $\dfrac{4}{5}$

7a. $\dfrac{5}{6x}$ 7b. $\dfrac{1}{3}$ 7c. $\dfrac{b^2}{4a}$

7d. $\dfrac{9p^2q}{r^3}$

Objective 1

1. yes   3. no; 4

Objective 2

5. $\dfrac{5}{7}$

Objective 3

7. $3 \cdot 5 \cdot 7$ 9. $2 \cdot 2 \cdot 2 \cdot 2 \cdot 2 \cdot 2 \cdot 5$

Objective 4

11. $\dfrac{3 \cdot 3 \cdot 7}{3 \cdot 5 \cdot 7}; \dfrac{3}{5}$

Objective 5

13. $\dfrac{3r}{s^2}$ 15. $\dfrac{4}{3x}$

Answers

4.3   Multiplying and Dividing Signed Fractions

Key Terms

1. reciprocals   2. indicator words   3. of

4. each

Now Try

1a. $\dfrac{40}{143}$ 1b. $-\dfrac{45}{88}$ 2a. $-\dfrac{2}{3}$

2b. $\dfrac{3}{20}$ 3. 15   4a. $\dfrac{14}{15}$

4b. $\dfrac{2c}{15}$ 5a. $\dfrac{1}{12}$ 5b. -80

5c. 4   5d. undefined   5e. 0

6a. $\dfrac{5a}{b}$ 6b. $\dfrac{10}{9b^2}$ 7a. $\dfrac{21}{22}$ m^2

7b. 72 servings

Objective 1

1. $-\dfrac{1}{7}$ 3. 45

Objective 2

5. $\dfrac{5}{2m}$

Objective 3

7. $-\dfrac{1}{24}$ 9. $\dfrac{6}{7}$

Objective 4

11. $6x$

Objective 5

13. $600

4.4   Adding and Subtracting Signed Fractions

Key Terms

1. unlike fractions   2. like fractions

3. least common denominator

Now Try

1a. $\dfrac{2}{3}$ 1b. $\dfrac{1}{2}$ 1c. $-\dfrac{3}{5}$

1d. $\dfrac{4}{a^2}$ 2a. 27   2b. 30

3a. 100   3b. 90   4a. $\dfrac{17}{18}$

4b. $-\dfrac{1}{6}$ 4c. $\dfrac{41}{72}$ 4d. $\dfrac{61}{7}$

5a. $\dfrac{10+3x}{15}$ 5b. $\dfrac{5x-64}{8x}$

Objective 1

1. $-\dfrac{2}{3}$ 3. $\dfrac{4}{y^2}$

Objective 2

5. 63

Objective 3

7. $\dfrac{3}{10}$ 9. $\dfrac{8}{15}$

Objective 4

11. $\dfrac{1+2x}{6}$

Answers

4.5   Problem Solving: Mixed Numbers and Estimating

Key Terms

1. mixed number   2. improper fraction

Now Try

1.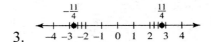

2. $\dfrac{35}{6}$

3a. $3\dfrac{3}{8}$ 3b. $4\dfrac{1}{9}$ 4a. 7

4b. 4   5a. 20; $17\dfrac{1}{2}$ 5b. 36; 35

6a. $\dfrac{4}{7}$; $\dfrac{2}{3}$ 6b. $\dfrac{7}{11}$; $\dfrac{5}{8}$ 7a. 6; $6\dfrac{5}{9}$

7b. 3; $3\dfrac{7}{12}$ 7c. 5; $4\dfrac{6}{7}$ 8a. 6 oz; $7\dfrac{25}{48}$ oz

8b. 10 ft; $10\dfrac{3}{40}$ ft

Objective 1

1.   3.

Objective 2

5. $-\dfrac{16}{9}$

Objective 3

7. 4; $4\dfrac{4}{21}$ 9. $2\dfrac{1}{2}$; $2\dfrac{2}{9}$

Objective 4

11. 4; $3\dfrac{5}{9}$

Objective 5

13. 12 m^2; $12\dfrac{1}{2}$ m^2

4.6 Exponents, Order of Operations, and Complex Fractions

Key Terms

1. order of operations   2. exponent   3. complex fraction

Now Try

1a. $-\dfrac{1}{64}$ 1b. $\dfrac{5}{28}$ 2a. $\dfrac{11}{30}$

2b. $-\dfrac{77}{81}$ 3a. $\dfrac{20}{21}$ 3b. $\dfrac{1}{25}$

Objective 1

1. $-\dfrac{1}{9}$ 3. $\dfrac{1}{36}$

Objective 2

5. $\dfrac{3}{4}$

Objective 3

7. -8 9. $\dfrac{25}{36}$

4.7 Problem Solving: Equations Containing Fractions

Key Terms

1. division property of equality

2. multiplication property of equality

Now Try

1a. 21   1b. -30 1c. $\dfrac{2}{5}$

2a. 30   2b. -21 3. 80 in. or 6 ft 8 in.

Objective 1

1. $b = -36$ 3. $h = \dfrac{3}{16}$

Objective 2

5. $r = -4$

Objective 3

7. 20 years old   9. 30 years old

Answers

4.8   Geometry Applications: Area and Volume

Key Terms

1. area   2. volume

Now Try

1a. $\dfrac{77}{128}$ in.^{2} 1b. 232.56 cm^2 2. 534 m^2

3a. 48 m^3 3b. $62\dfrac{1}{2}$ in.^{3} 4. 25.8 ft^3

Objective 1

1. $P = 66$ m; $A = 200$ m^2 3. 1940 yd^2

Objective 2

5. $166\dfrac{3}{8}$ ft^3

Objective 3

7. 196 ft^3 9. $139\dfrac{1}{2}$ m^3

Chapter 5 RATIONAL NUMBERS: POSITIVE AND NEGATIVE DECIMALS

5.1 Reading and Writing Decimals

Key Terms

1. decimals
2. place value
3. decimal point

Now Try

1a. 0.6
1b. 0.05
1c. 0.37

1d. 0.007
1e. 0.049
1f. 0.518

2a. 8 hundreds; 6 tens; 2 ones; 9 tenths; 3 hundredths

2b. 0 ones; 0 tenths; 0 hundredths; 7 thousandths; 6 ten-thousandths; 9 hundred-thousandths

3a. nine tenths
3b. fifty-three hundredths
3c. seven hundredths

3d. five hundred two thousandths

3e. four hundred sixty-nine ten-thousandths
4a. four and seven tenths

4b. eighteen and nine thousandths
4c. eight-two ten-thousandths

4d. fifty-seven and nine hundred six thousandths
5a. $\dfrac{27}{100}$

5b. $\dfrac{303}{1000}$
5c. $3\dfrac{2636}{10,000}$
6a. $\dfrac{1}{5}$

6b. $\dfrac{7}{20}$
6c. $6\dfrac{1}{25}$
6d. $562\dfrac{101}{2500}$

Objective 1

1. $\dfrac{8}{10}$; 0.8; eight tenths
3. $\dfrac{58}{100}$; 0.58; fifty-eight hundredths

Objective 2

5. 7; 1

Objective 3

7. eight hundredths
9. ninety seven and eight thousandths

Objective 4

11. $3\dfrac{3}{5}$

5.2   Rounding Decimal Numbers

Key Terms

1. decimal places   2. rounding

Now Try

1. 43.803	2a. 0.80	2b. 7.78
2c. 22.040	2d. 0.6	3a. $2.08
3b. $425.10	4a. $38	4b. $307
4c. $881	4d. $6860	4e. $1

Objective 1

1. up

Objective 2

3. 489.8   5. 989.990

Objective 3

7. $11,840

5.3   Adding and Subtracting Signed Decimal Numbers

Key Terms

1. front end rounding   2. estimating

Now Try

1a. 15.630	1b. 21.709	2a. 36.844
2b. 1013.931	3a. 11.612	3b. 10.57
4a. 4.654	4b. 3.26	4c. 0.501
5a. −38.4	5b. −10.586	6a. −11.74
6b. 3.31	6c. 16.54	7a. 14; 13.847
7b. $50; $41.55	7c. −11; −11.097	

Objective 1

1. 92.49   3. 72.453

Objective 2

5. 178

Objective 3

7. 680; 676.60   9. −15; −14.562

5.4 Multiplying Signed Decimal Numbers

Key Terms

1. factor   2. decimal places   3. product

Now Try

1. −10.793   2. 0.00186   3. 100; 112.8229

Objective 1

1. −90.71   3. 0.0037

Objective 2

5. 600; 756.6478

5.5 Dividing Signed Decimal Numbers

Key Terms

1. dividend   2. repeating decimal   3. quotient

4. divisor

Now Try

1a. −1.413   1b. 15.03   2. 20.178

3. 16.791   4a. 20,410   4b. 1.78

5. 40; 37.8   6a. 12.18   6b. 22.62

6c. −0.233

Objective 1

1. 6.96   3. 2.359

Objective 2

5. 128.25

Objective 3

7. reasonable   9. reasonable

Objective 4

11. 53.24

5.6   Fractions and Decimals

Key Terms

1. equivalent   2. numerator   3. denominator

4. mixed number

Now Try

1a. 0.0625   1b. 4.375   2. 0.417

3a. >   3b. =   3c. <

3d. =   4a. 0.3, 0.3057, 0.307   4b. $\frac{1}{5}$, $\frac{2}{9}$, 0.23, $\frac{3}{13}$

Objective 1

1. 0.125   3. 19.708

Objective 2

5. $\frac{3}{11}$, 0.29, $\frac{1}{3}$

5.7   Problem Solving with Statistics: Mean, Median, and Mode

Key Terms

1. mode   2. weighted mean   3. mean

4. median

Now Try

1. 272   2. 52.7   3. 40.2

4. 2.5   5. 18   6. 0.06

7a. 3   7b. 13 and 16   7c. no mode

Objective 1

1. 60.3   3. 273.1

Objective 2

5. 4.7

Objective 3

7. 232   9. 239.5

Objective 4

11. 24, 35, 39

5.8   Geometry Applications: Pythagorean Theorem and Square Roots

Key Terms

1. right triangle   2. hypotenuse   3. legs

Now Try

1a. 4.47   1b. 9.59   1c. 11.58

2a. 9.8 in.   2b. 12 km   3. 9.5 ft

Objective 1

1. 4.123   3. 10.100

Objective 2

5. 2.2 cm

Objective 3

7. 10.3 ft   9. 8 ft

5.9   Problem Solving: Equations Containing Decimals

Key Terms

1. division property of equality

2. addition property of equality

Now Try

1a. $w = -13.9$ 1b. $x = 19.6$ 2a. $x = 3.05$

2b. $t = -8.5$ 3a. $b = -1.6$ 3b. $x = 0.53$

4. 32 min

Objective 1

1. $n = -4.6$ 3. $h = 7.7$

Objective 2

5. $r = 2.5$

Objective 3

7. $y = -15$ 9. $x = 29.4$

Objective 4

11. 30 mg

403

5.10 Geometry Applications: Circles, Cylinders, and Surface Area

Key Terms

1. radius
2. circumference
3. circle
4. π (pi)
5. surface area
6. diameter

Now Try

1a. 86 m
1b. 29.5 in.
2a. 138.2 yd

2b. 23.2 m
3a. 43.0 m^2
3b. 1519.8 yd^2

4. 127.2 ft^2
5. $12.56
6. $1.96

7a. 490.6 cm^3
7b. 678.2 cm^3

8. $V = 7956$ mm^3; $S = 2670$ mm^2

9. $V \approx 113.0$ ft^3; $S \approx 138.2$ ft^2

Objective 1

1. 4 ft
3. $6\frac{1}{4}$ yd

Objective 2

5. 28.3 yd

Objective 3

7. 22.3 yd^2
9. 1061.3 m^2

Objective 4

11. 6358.5 cm^3

Objective 5

13. 558 in.^{2}
15. 85.7 in.^{2}

Objective 6

17. 183.6 ft^2

Chapter 6 RATIO, PROPORTION, AND LINE/ANGLE/TRIANGLE RELATIONSHIPS

6.1   Ratios

Key Terms

1. ratio

2. numerator; denominator

Now Try

1a. $\dfrac{17}{8}$

1b. $\dfrac{8}{5}$

2a. $\dfrac{2}{3}$

2b. $\dfrac{2}{1}$

2c. $\dfrac{2}{3}$

3. $\dfrac{3}{8}$

4a. $\dfrac{19}{9}$

4b. $\dfrac{22}{17}$

5. $\dfrac{5}{7}$

Objective 1

1. $\dfrac{25}{19}$

3. $\dfrac{1}{4}$

Objective 2

5. $\dfrac{9}{2}$

Objective 3

7. $\dfrac{2}{7}$

9. $\dfrac{5}{6}$

Answers

6.2   Rates

Key Terms

1. unit rate   2. cost per unit   3. rate

Now Try

1a. $\dfrac{1 \text{ dollar}}{5 \text{ pages}}$ 1b. $\dfrac{15 \text{ strokes}}{1 \text{ minute}}$ 1c. $\dfrac{33 \text{ strawberries}}{2 \text{ cakes}}$

2a. 28 miles/gallon   2b. $1.26/pound   2c. $145/day

3. 5 pints at $1.70/pint   4a. eight-pack   4b. Brand T

Objective 1

1. $\dfrac{7 \text{ pills}}{1 \text{ patient}}$ 3. $\dfrac{32 \text{ pages}}{1 \text{ chapter}}$

Objective 2

5. $225/pound

Objective 3

7. 24 ounces for $2.08   9. 5 cans for $2.75

6.3   Proportions

Key Terms

1. proportion   2. cross products   3. ratio

Now Try

1a. $\dfrac{24}{17}=\dfrac{72}{51}$

1b. $\dfrac{\$10}{7\text{ cans}}=\dfrac{\$60}{42\text{ cans}}$

2a. $\dfrac{9}{7}\neq\dfrac{4}{3}$, false

2b. $\dfrac{1}{3}=\dfrac{1}{3}$, true

3a. $306=306$, true

3b. $32\neq 35$, false

4a. 32

4b. 10.67

5a. $2\dfrac{3}{5}$

5b. 6.4

Objective 1

1. $\dfrac{50}{8}=\dfrac{75}{12}$

3. $\dfrac{3}{33}=\dfrac{12}{132}$

Objective 2

5. $\dfrac{6}{5}=\dfrac{6}{5}$; true

Objective 3

7. 36

9. 21

6.4   Problem Solving with Proportions

Key Terms

1. ratio   2. rate

Now Try

1. $\dfrac{23\text{ hr}}{4\text{ apt}}=\dfrac{x\text{ hr}}{16\text{ apt}}$; $x=92$ hr

2. $\dfrac{4}{5}=\dfrac{x}{540}$; $x=432$ people

Objective 1

1. $108

3. approximately 333 deer

6.5   Geometry: Lines and Angles

Key Terms

1. ray
2. perpendicular lines
3. obtuse angle
4. point
5. angle
6. line
7. acute angle
8. degrees
9. parallel lines
10. line segment
11. intersecting lines
12. right angle
13. vertical angles
14. congruent angles
15. supplementary angles
16. complementary angles
17. alternate interior angles
18. corresponding angles
19. straight angle

Now Try

1a. line segment, $\overline{EF}$
1b. ray, $\overrightarrow{CD}$
1c. line, $\overleftrightarrow{RS}$

2a. intersecting
2b. parallel
3. $\angle VSW$ or $\angle WSV$

4a. straight
4b. obtuse
4c. acute

4d. right
5a. intersecting
5b. perpendicular

6. $\angle SRT$ and $\angle TRU$; $\angle URV$ and $\angle VRW$
7a. $18°$

7b. $78°$

8. $\angle OQP$ and $\angle PQN$; $\angle RST$ and $\angle BMC$; $\angle OQP$ and $\angle BMC$, $\angle PQN$ and $\angle RST$

9a. $98°$
9b. $12°$

10. $\angle RPS \cong \angle QPT$ and $\angle QPR \cong \angle TPS$

11. $\angle MKN$ and $\angle QKP$; $\angle MKQ$ and $\angle NKP$
12a. $33°$

12b. $105°$
12c. $42°$
12d. $42°$

13a. $\angle 5$ and $\angle 6$; $\angle 1$ and $\angle 2$; $\angle 7$ and $\angle 8$; $\angle 3$ and $\angle 4$

13b. $\angle 3$ and $\angle 6$; $\angle 7$ and $\angle 2$

14. $m\angle 1 = 55°$; $m\angle 2 = 55°$; $m\angle 3 = 125°$; $m\angle 4 = 125°$
$m\angle 5 = 125°$; $m\angle 6 = 125°$; $m\angle 7 = 55°$; $m\angle 8 = 55°$

Objective 1

1. ray, $\overrightarrow{CB}$ 3. line segment, $\overline{KL}$

Objective 2

5. parallel

Objective 3

7. $\angle COD$ 9. $\angle MON$

Objective 4

11. obtuse

Objective 5

13. perpendicular   15. intersecting

Objective 6

17. 164°

Objective 7

19. $\angle CAD \cong \angle BAE$ 21. 73° and 107°

Objective 8

23. corresponding angles: $\angle 1$ and $\angle 5$; $\angle 2$ and $\angle 6$; $\angle 3$ and $\angle 7$; $\angle 4$ and $\angle 8$
alternate interior angles: $\angle 4$ and $\angle 6$; $\angle 3$ and $\angle 5$
$m\angle 1 = 37°$; $m\angle 2 = 143°$; $m\angle 3 = 37°$; $m\angle 4 = 143°$
$m\angle 5 = 37°$; $m\angle 6 = 143°$; $m\angle 7 = 37°$; $m\angle 8 = 143°$

6.6   Geometry Applications: Congruent and Similar Triangles

Key Terms

1. congruent triangles   2. congruent figures   3. similar triangles

4. similar figures

Now Try

1. $\angle 1$ and $\angle 5$; $\angle 2$ and $\angle 4$; $\angle 3$ and $\angle 6$; $\overline{ST}$ and $\overline{VX}$; $\overline{RS}$ and $\overline{WV}$; $\overline{RT}$ and $\overline{WX}$

2a. SAS   2b. ASA   2c. SSS

3. $\overline{AB}$ and $\overline{PQ}$; $\overline{AC}$ and $\overline{PR}$; $\overline{BC}$ and $\overline{QR}$;
$\angle A$ and $\angle P$; $\angle B$ and $\angle Q$; $\angle C$ and $\angle R$

4. 12.75 yd   5. 24   6. 21 ft

Objective 1

1. $\angle 1$ and $\angle 4$; $\angle 2$ and $\angle 5$; $\angle 3$ and $\angle 6$; $\overline{DF}$ and $\overline{XZ}$; $\overline{DE}$ and $\overline{XY}$; $\overline{EF}$ and $\overline{YZ}$

Objective 2

3. ASA

Objective 3

5. $\overline{PN}$ and $\overline{SR}$; $\overline{NM}$ and $\overline{RQ}$; $\overline{MP}$ and $\overline{QS}$;
$\angle P$ and $\angle S$; $\angle N$ and $\angle R$; $\angle M$ and $\angle Q$

7. $\overline{HK}$ and $\overline{RS}$; $\overline{GH}$ and $\overline{TR}$; $\overline{GK}$ and $\overline{TS}$;
$\angle H$ and $\angle R$; $\angle G$ and $\angle T$; $\angle K$ and $\angle S$

Objective 4

9. $m = 90$; $r = 21$

Objective 5

11. 36 m   13. 10 m

Chapter 7 PERCENTS

7.1 The Basics of Percent
Key Terms

1. ratio 　　　　2. percent 　　　　3. decimals

Now Try

1a. 10%　　　　1b. 18%　　　　1c. 85%

2a. 0.23　　　　2b. 0.08　　　　2c. 0.153

2d. 5.00　　　　2e. 3.02　　　　3a. 0.48

3b. 2.40 or 2.4　　3c. 0.025　　　3d. 0.008

4a. 43%　　　　4b. 230%　　　　4c. 75.1%

5a. $\frac{3}{10}$　　　5b. $\frac{39}{50}$　　　5c. $1\frac{3}{4}$

6a. $\frac{109}{250}$　　6b. $\frac{2}{9}$　　　7a. 15%

7b. $93\frac{3}{4}\%$　　7c. $5\frac{5}{9}\%$ exact, or 5.6% rounded

8a. $6.15　　　8b. $23\frac{1}{2}$ days　　9a. $3.65

9b. $9\frac{1}{2}$ pages

Objective 1

1. 68%　　　　3. 45%

Objective 2

5. 3.10

Objective 3

7. 20%　　　　9. 493%

Objective 4

11. $\frac{11}{60}$

Objective 5

13. 94%　　　　15. 85.3%

Objective 6

17. $520

7.2 The Percent Proportion

Key Terms

1. whole   2. part   3. percent proportion

Now Try

1a. 59   1b. 7.75   1c. unknown

2a. 650   2b. unknown   2c. 3280

3a. unknown   3b. 950   3c. 2650

4. 2470   5. 4%   6. 150 cars

7a. 220%   7b. $94

Objective 1

1. 250%; unknown; $50   3. $7\frac{3}{4}\%$; $895; unknown

Objective 2

5. 141.0%

7.3 The Percent Equation

Key Terms

1. percent   2. percent equation

Now Try

1a. $50   1b. 15 hr   1c. 25 pounds

2a. $93.20   2b. 69.3 miles   3a. $5.49

3b. 0.307 mile   4a. $275.90   4b. 108 miles

5a. 220%   5b. 33%   6a. 300 tons

6b. 78

Objective 1

1. 25 minutes   3. $2000

Objective 2

5. 492 televisions

Objective 3

7. 75%   9. 400 magazines

7.4 Problem Solving with Percent

Key Terms

1. percent equation   2. percent of increase or decrease

3. percent proportion

Now Try

1. 66 points   2. 294 drivers   3. 23.3%

4. 165%   5. 77 micrograms   6. 125%

7. 28.3%

Objective 1

1. 782 members   3. 17%

Objective 2

5. 28.6%

7.5 Consumer Applications: Sales Tax, Tips, Discounts, and Simple Interest

Key Terms

1. interest rate   2. interest   3. interest formula

4. simple interest   5. principal   6. sales tax

7. discount   8. tax rate

Now Try

1. Tax: $6.23; Total: $95.23   2. 12%

3a. Est: $3; Exact: $3.56   3b. Est: $4; Exact: $4.75   4. $9.98

5. $4; $84   6. $124; $744   7. $29.75; $869.75

Objective 1

1. $22.75; $372.75   3. 5%

Objective 2

5. $9.00; $9.58; $12.00; $12.77

Objective 3

7. $30; $170   9. $36.18

Objective 4

11. $195; $975

Chapter 8 MEASUREMENT

8.1   Problem Solving with U.S. Customary Measurements

Key Terms

1. metric system   2. unit fractions

3. U.S. measurement units

Now Try

1a. 1 lb   1b. 4 qt   2a. 210 inches

2b. 56 ounces   2c. 1.25 or $1\frac{1}{4}$ minutes   2d. $6.3\overline{3}$ or $6\frac{1}{3}$ hours

3a. 432 inches   3b. $\frac{1}{3}$ ft   4a. 21,120 feet

4b. $1\frac{1}{2}$ tons   4c. 13 quarts   5a. $\frac{3}{4}$ gal

5b. 7200 minutes   6a. $6.36 per pound   6b. 3.5 or $3\frac{1}{2}$ tons

Objective 1

1. 2000   3. 8

Objective 2

5. 5 gallons

Objective 3

7. 3.5 or $3\frac{1}{2}$ gallons   9. 3.75 or $3\frac{3}{4}$ pounds

Objective 4

11. $26\frac{1}{4}$ qt

8.2   The Metric System—Length

Key Terms

1. prefix   2. metric conversion line   3. meter

Now Try

1a. mm   1b. km   1c. cm

2a. 5.4 km   2b. 76 mm   3a. 675 mm

3b. 0.9865 km   3c. 431 cm   4a. 23,000 mm

4b. 0.042 m   4c. 0.95 km

Objective 1

1. m   3. km

Objective 2

5. 0.45 km

Objective 3

7. 19.4 mm   9. 0.000035 cm

8.3   The Metric System—Capacity and Weight (Mass)

Key Terms

1. gram   2. liter

Now Try

1a. 5 mL   1b. 100 L   2a. 0.973 L

2b. 3850 mL   3a. 48 kg   3b. 590 g

3c. 3 mg   4a. 3720 mg   4b. 0.084 kg

5a. 125 mL   5b. 1.19 kg   5c. 400 m

Objective 1

1. mL   3. mL

Objective 2

5. 836,000 L

Objective 3

7. mg   9. g

Objective 4

11. 760 g

Objective 5

13. L   15. cm

8.4   Problem Solving with Metric Measurement

Key Terms

1. gram

2. meter

3. liter

Now Try

1. 1.54 kg

2. 200 g

3. 5600 mL

Objective 1

1. $12.99

3. 12 pills

8.5   Metric–U.S. Measurement Conversions and Temperature

Key Terms

1. Celsius

2. Fahrenheit

Now Try

1. 42.3 ft

2a. 21.3 lb

2b. 75.8 L

3a. 65°C

3b. 4°C

4. 50°C

5. 302°F

Objective 1

1. 468.5 km

3. 737.1 g

Objective 2

5. 37°C

Objective 3

7. 17°C

9. 204°C

Chapter 9 GRAPHS AND GRAPHING

9.1 Problem Solving with Tables and Pictographs

Key Terms

1. pictograph
2. table

Now Try

1a. 93% 1b. Southwest and United 1c. 80%

2a. $64.50 2b. $77.50 3a. 45 million

3b. 45 million

Objective 1

1. 177 calories 3. 486 calories

Objective 2

5. 3 million people

Answers

9.2   Reading and Constructing Circle Graphs

Key Terms

1. circle graph   2. protractor

1. $\dfrac{3}{58}$ 2. $\dfrac{5}{9}$ 3. $142,500

4a. mysteries: 108°; biographies: 54°; cookbooks: 36°; romance novels: 90°; science: 54°; business: 18°

4b.

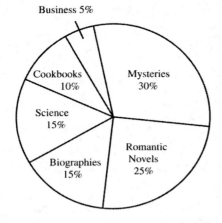

Objective 1

1. $10,400   3. $400

Objective 2

5. $\dfrac{1000}{1800}$ or $\dfrac{5}{9}$

Objective 3

7. parts: 5%; hand tools: 20%; bench tools: 25%; brass fittings: 35%; hardware: 15%

9.

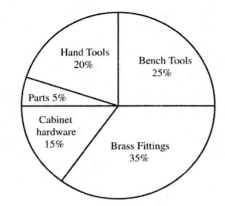

9.3 Bar Graphs and Line Graphs

Key Terms

1. line graph

2. double-bar graph

3. bar graph

4. comparison line graph

Now Try

1. 1800 students

2a. 250 female seniors

2b. 500 male freshmen

3a. $60

3b. December

4a. $1,500,000

4b. $4,000,000

4c. $1,000,000; 33%

Objective 1

1. 1600 students

3. 400 students

Objective 2

5. sophomores

Objective 3

7. September

9. $30

Objective 4

11. $2,000,000

Answers

9.4   The Rectangular Coordinate System

Key Terms

1. ordered pair   2. horizontal axis; *x*-axis   3. vertical axis; *y*-axis

4. paired data   5. coordinate system   6. quadrants

7. origin   8. coordinates

Now Try

1a. and 1b.

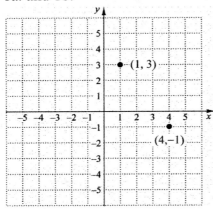

2a. and 2b.

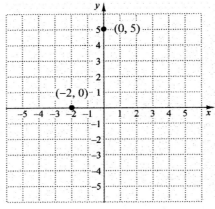

3a. (–4, –4)   3b. (1, –3)   3c. (2, 4)

3d. (0, –5)   4a. Quadrant I   4b. Quadrant II

4c. not in any quadrant

Objective 1

1.

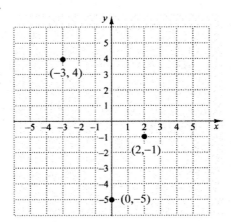

Objective 2

3. Quadrant IV   5. none

9.5   Introduction to Graphing Linear Equations

Key Terms

1. graph a linear equation

2. slope

Now Try

1.

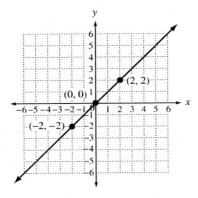

2.

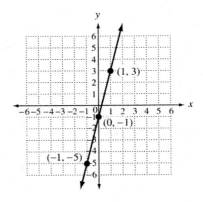

3.

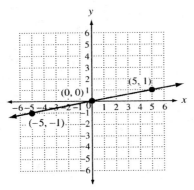

4.

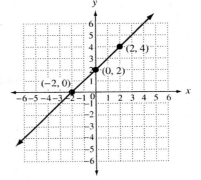

5. positive; increases

Objective 1

1.

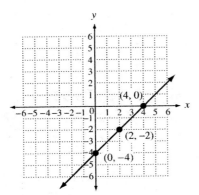

Objective 2

3. positive

Answers

Chapter 10 EXPONENTS AND POLYNOMIALS

10.1   The Product Rule and Power Rules for Exponents

Key Terms

1. power   2. exponential expression   3. base

Now Try

1. 4^5; 1024   2a. base: 2; exponent: 6; value: 64

2b. base:2; exponent: 6; value: -64

2c. base: -2; exponent: 6; value 64   3a. 9^{13}

3b. $(-6)^6$ 3c. x^6 3d. m^{27}

3e. 500   3f. 108   4. $18x^{11}$

5a. 4^{15} 5b. 7^8 5c. x^{49}

5d. n^{30} 6a. $64a^3b^3$ 6b. $7p^3q^3$

6c. $200x^8y^6$ 7a. $\dfrac{125}{216}$ 7b. $\dfrac{p^6}{q^6}$

Objective 1

1. $\left(\dfrac{1}{3}\right)^5$; $\dfrac{1}{243}$ 3. -6561; base: 3; exponent: 8

Objective 2

5. $8c^{15}$

Objective 3

7. 7^{12} 9. 2^{21}

Objective 4

11. $27a^{12}b^3$

Objective 5

13. $-\dfrac{8x^3}{125}$ 15. $-\dfrac{128a^7}{b^{14}}$

10.2   Integer Exponents and the Quotient Rule

Key Terms

1. power rule for exponents

2. base; exponent

3. product rule for exponents

Now Try

1a. 1

1b. 1

1c. −1

1d. 1

1e. 88

1f. 1

2a. $\dfrac{1}{64}$

2b. $\dfrac{1}{27}$

2c. $\dfrac{1}{8}$

2d. $\dfrac{1}{p^4}$

3a. 9

3b. $\dfrac{1}{16}$

3c. 64

3d. z^{10}

4a. 10^1 or 10

4b. $\dfrac{1}{10^{14}}$

4c. $\dfrac{1}{y^8}$

Objective 1

1. −1

3. 0

Objective 2

5. $\dfrac{1}{n^9}$

Objective 3

7. 13^7

9. $\dfrac{1}{z^{36}}$

Objective 4

11. p^{12}

Answers

10.3   An Application of Exponents: Scientific Notation
Key Terms

1. scientific notation   2. power rule   3. quotient rule

Now Try

1a. 6.8×10^8

1b. 4.771×10^{10}

1c. 9.991×10^0

1d. 4.63×10^{-2}

1e. -8.48×10^{-4}

2a. 835,000

2b. 27,960,000

2c. −0.000164

3a. 2.7×10^8, or 270,000,000

3b. 3×10^{-8}, or 0.00000003

4. 4.86×10^{19} atoms

5. 5.45×10^3 kg/m^3

Objective 1

1. 2.3651×10^4

3. -2.208×10^{-4}

Objective 2

5. 0.0064

Objective 3

7. 2.53×10^2

Objective 4

9. 2.23×10^9 pretzels or 2,230,000,000 pretzels

11. 3.34×10^{12} kg/m^3 or 3,340,000,000,000 kg/m^3

10.4   Adding and Subtracting Polynomials

Key Terms

1. degree of a term
2. descending powers
3. trinomial
4. polynomial
5. monomial
6. degree of a polynomial
7. binomial

Now Try

1a. $5x^4$

1b. $-8x^7$

1c. $15m^3 + 29m^2$

1d. $5p^2q$

2a. $8x^3 + 4x^2 + 6$; degree 3; trinomial

2b. $4x^5$; degree 5; monomial

3. 1285; 1357

4a. $5x^3 - 5x^2 + 4x$

4b. $9x^4 + 7x^2 + 7x - 6$

5a. $4x^3 + x + 12$

5b. $7x^3 + 8x^2 - 14x - 1$

6a. $2x + 2$

6b. $-11x^3 - 7x + 1$

Objective 1

1. $-3z^3$

3. $8c^3 - 8c^2 - 6c + 6$

Objective 2

5. $-5.7d^8 - 1.1d^5 + 3.2d^3 - d^2$; degree 8; none of these

Objective 3

7. a. -51; b. 74

9. a. 29; b. 39

Objective 4

11. $4x^2 + 6x - 18$

Objective 5

13. $-8w^3 + 21w^2 - 15$

15. $3x^3 + 7x^2 + 2$

Answers

10.5 Multiplying Polynomials: An Introduction

Key Terms

1. monomial

2. distributive property

3. trinomial

4. polynomial

5. binomial

Now Try

1a. $32x^4 + 64x^3$

1b. $-35m^8 + 42m^7 - 28m^6 + 7m^5$

2a. $x^2 + 3x - 54$

2b. $4x^7 - 2x^5 + 37x^4 - 18x^2 + 9x$

3. $28x^4 - 33x^3 + 51x^2 + 17x - 15$

Objective 1

1. $35z^4 + 14z$

3. $-6y^5 - 9y^4 + 12y^3 - 33y^2$

Objective 2

5. $6m^5 + 4m^4 - 5m^3 + 2m^2 - 4m$

Chapter R WHOLE NUMBERS REVIEW

R.1   Adding Whole Numbers
Key Terms

1. commutative property of addition

2. addends

3. addition

4. associative property of addition

5. regrouping

6. sum (total)

Now Try

1a. $7 + 2 = 9$ and $2 + 7 = 9$

1b. $5 + 9 = 14$ and $9 + 5 = 14$

1c. $3 + 3 = 6$

2. 29

3. 788

4. 65

5. 11,179

6. 33 miles

7. 44 miles

8. incorrect; 1948

Objective 1

1. 12

Objective 2

3. 31

Objective 3

5. 987

Objective 4

7. 1041

9. 10,415

Objective 5

11. 44 miles

Objective 6

13. incorrect; 8612

R.2   Subtracting Whole Numbers

Key Terms

1. minuend
2. regrouping
3. subtrahend
4. difference

Now Try

1a. $13 - 8 = 5$ or $13 - 5 = 8$

1b. $12 - 9 = 3$ or $12 - 3 = 9$

2a. $9 = 7 + 2$ or $9 = 2 + 7$

2b. $320 = 190 + 130$ or $320 = 130 + 190$

3. minuend: 87; subtrahend: 12; difference: 75

4a. 43

4b. 111

4c. 6210

5a. correct

5b. incorrect;222

5c. incorrect; 1302

6. 39

7. 488

8. 2788

9. 5003

10a. correct

10b. incorrect; 378

11. 95 passengers

Objective 1

1. $187 - 38 = 149$; $187 - 149 = 38$

3. $785 + 426 = 1211$

Objective 2

5. minuend: 35; subtrahend: 9; difference: 24

Objective 3

7. 6012

Objective 4

9. incorrect; 153

Objective 5

11. 198

13. 6703

Objective 6

15. $263

R.3 Multiplying Whole Numbers

Key Terms

1. commutative property of multiplication
2. factors
3. associative property of multiplication
4. multiple
5. product
6. partial products
7. chain multiplication problem

Now Try

1a. 42	1b. 0	1c. 16
2. 40	3a. 258	3b. 3969
4a. 460	4b. 70,800	4c. 519,000
5a. 17,800	5b. 8400	6a. 2880
6b. 28,116	6c. 8745	7a. 89,358
7b. 32,388,875	8. $5372	

Objective 1

1. factors: 9, 12; 108   3. 72

Objective 2

5. 84

Objective 3

7. 216   9. 180,054

Objective 4

11. 20,000

Objective 5

13. 4416   15. 1,644,648

Objective 6

17. $9728

R.4 Dividing Whole Numbers

Key Terms

1. remainder   2. dividend   3. quotient

4. divisor   5. short division

Now Try

1a. $4\overline{)28}^{\,7}$ or $\dfrac{28}{4}=7$ 1b. $72\div 9=8$ or $9\overline{)72}^{\,8}$ 1c. $32\div 8=4$ or $\dfrac{32}{8}=4$

2a. dividend: 63; divisor: 9; quotient: 7   2b. dividend: 56; divisor: 8; quotient: 7

2c. dividend: 54; divisor: 6; quotient: 9   3a. 0

3b. 0   3c. 0   3d. 0

4a. $4\cdot 9=36$ or $9\cdot 4=36$ 4b. $3\cdot 6=18$ or $6\cdot 3=18$ 4c. $8\cdot 5=40$ or $5\cdot 8=40$

5a. undefined   5b. undefined   5c. undefined

6a. 1   6b. 1   6c. 1

7. 23   8. 74 R 1   9. 261 R 5

10a. correct   10b. incorrect; 78 R1   11a. Yes

11b. No   12a. Yes   12b. No

13a. Yes   13b. Yes   13c. No

14a. Yes   14b. No

Objective 1

1. $3\overline{)15}^{\,5}$; $\dfrac{15}{3}=5$

Objective 2

3. dividend: 63; divisor: 7; quotient: 9   5. dividend: 44; divisor: 11; quotient: 4

Objective 3

7. 0

Objective 4

9. undefined   11. undefined

Objective 5

13. 1

Objective 6

15. 72 R 3

Objective 7

17. incorrect; 1522 R 5   19. incorrect; 6050

Objective 8

21. 2: yes; 3: no; 5: no; 10: no

R.5   Long Division

Key Terms

1. dividend   2. remainder   3. divisor

4. quotient   5. long division

Now Try

1. 73   2. 38 R 55   3. 209 R20

4a. 5   4b. 47   4c. 606

5a. 135   5b. 2006   6. incorrect; 37 R62

Objective 1

1. 77   3. 654 R22

Objective 2

5. 13

Objective 3

7. correct   9. incorrect; 460 R3

431